Advances in Intelligent Systems and Computing

Volume 1441

AF868654

The series "Advances in Intelligent Systems and Computing" contains publications on theory, applications, and design methods of Intelligent Systems and Intelligent Computing. Virtually all disciplines such as engineering, natural sciences, computer and information science, ICT, economics, business, e-commerce, environment, healthcare, life science are covered. The list of topics spans all the areas of modern intelligent systems and computing such as: computational intelligence, soft computing including neural networks, fuzzy systems, evolutionary computing and the fusion of these paradigms, social intelligence, ambient intelligence, computational neuroscience, artificial life, virtual worlds and society, cognitive science and systems, Perception and Vision, DNA and immune based systems, self-organizing and adaptive systems, e-Learning and teaching, human-centered and human-centric computing, recommender systems, intelligent control, robotics and mechatronics including human-machine teaming, knowledge-based paradigms, learning paradigms, machine ethics, intelligent data analysis, knowledge management, intelligent agents, intelligent decision making and support, intelligent network security, trust management, interactive entertainment, Web intelligence and multimedia.

The publications within "Advances in Intelligent Systems and Computing" are primarily proceedings of important conferences, symposia and congresses. They cover significant recent developments in the field, both of a foundational and applicable character. An important characteristic feature of the series is the short publication time and world-wide distribution. This permits a rapid and broad dissemination of research results.

Indexed by DBLP, INSPEC, WTI Frankfurt eG, zbMATH, Japanese Science and Technology Agency (JST).

All books published in the series are submitted for consideration in Web of Science.

For proposals from Asia please contact Aninda Bose (aninda.bose@springer.com).

Mohamed Abd Elaziz · Mohamed Medhat Gaber ·
Shaker El-Sappagh · Mohammed A. A. Al-qaness ·
Ahmed A. Ewees
Editors

International Conference on Artificial Intelligence Science and Applications (CAISA)

Springer

Editors
Mohamed Abd Elaziz
Galala University
Al Galala, Egypt

Shaker El-Sappagh
Galala University
Al Galala, Egypt

Ahmed A. Ewees
Damietta University
Damietta, Egypt

Mohamed Medhat Gaber
Faculty of Computer Science
and Engineering
Galala University
Al Galala, Egypt

Birmingham City University
Birmingham, United Kingdom

Mohammed A. A. Al-qaness
College of Physics and Electronic
Information Engineering
Zhejiang Normal University
Jinhua, China

ISSN 2194-5357 ISSN 2194-5365 (electronic)
Advances in Intelligent Systems and Computing
ISBN 978-3-031-28105-1 ISBN 978-3-031-28106-8 (eBook)
https://doi.org/10.1007/978-3-031-28106-8

This Springer imprint is published by the registered company Springer Nature Switzerland AG
The registered company address is: Gewerbestrasse 11, 6330 Cham, Switzerland

Contents

A Hybrid Model for Predicting Road Accident Severity in Senegal 1
Yoro Dia, Lamine Faty, Aba Diop, Ousmane Sall, and Tony Tona Landu

An Efficient Machine Learning Algorithm for Breast Cancer Prediction 13
Yousif A. Al Haj, Marwan M. Al Falah, Abdullah M. Al-Arshy, Khadeja M. Al-Nashad, Zain Alabedeen A. Al-Nomi, Badr A. Al-Badawi, and Mustafa S. Al-Khayat

A New Similarity Measure for Multi Criteria Recommender System ... 29
Rizwan Abbas, Qaisar Abbas, Gehad Abdullah Amran, Abdulaziz Ali, Majed Hassan Almusali, Ali A. AL-Bakhrani, and Mohammed A. A. Al-qaness

Speeding Up and Enhancing the Hyperspectral Images Classification 53
Dalal AL-Alimi, Mohammed A. A. Al-qaness, and Zhihua Cai

A Healthcare System Based on Fog Computing 63
Maha Abdulaziz Alhazzani, Samia Allaoua Chelloug, Reema Abdulaziz Alomari, Maha Khalid Alshammari, and Reem Shaya Alqahtani

An Improved Arithmetic Optimization Algorithm with Differential Evolution and Chaotic Local Search 81
Aminu Onimisi Abdulsalami, Mohamed Abd Elaziz, Yousif A. Al Haj, and Shengwu Xiong

Services Management in the Digital Era—The Cloud Computing Perspective 97
Aaqif Afzaal Abbasi and Mohammad A. A. Al-qaness

Trip Recommendation Using Location-Based Social Network: A Review 107
Rizwan Abbas, Irshad Hussain, Gehad Abdullah Amran, Sultan Trahib Alotaibi, Ali A. AL-Bakhrani, Esmail Almosharea, and Mohammed A. A. Al-qaness

A Multimodal Spam Filtering System for Multimedia Messaging Service 121
Insaf Kraidia, Afifa Ghenai, and Nadia Zeghib

Optimized Neural Network for Evaluation Cisplatin Role in Neoplastic Treatment 133
Ahmed T. Sahlol, Ahmed A. Ewees, and Yasmine S. Moemen

A Hybrid Model for Predicting Road Accident Severity in Senegal

Yoro Dia, Lamine Faty, Aba Diop, Ousmane Sall, and Tony Tona Landu

Abstract Road transport system is one of the key elements to ensure the life of the population and the normal functioning of production and development processes. In this paper, we propose a study based on Artificial Intelligence through the Machine Learning approach for contribution to the improvement of Senegalese road safety. In previous works, we studied the severity of accidents by comparing the performance of Supervised Learning algorithms such as *Logistic Regression*, Random Forest, *K Nearest Neighbors*, *SVM*, and *Naive Bayesian classifier*. The best performance was obtained by Random Forest with a percentage performance of 85.60%. The performance of the other classifiers was significantly lower, with a percentage performance not exceeding 84%. In this study, we went further with the Random Forest algorithm in understanding the parameterization mechanisms. On the other hand, we also implemented *Random Forest Features Importance* (RFFI) to detect the features that have more influence on the model. We put forth a hybrid model that combines random forest (RF) and random search (RS) optimization techniques. In the suggested model, RF-RS, RF serves as the fundamental predictive model while RS is employed to optimize RF's hyperparameters. The experimental results demonstrate that RF-RS outperforms traditional algorithms in terms of performance.

Keywords Road accident · Random forest · Random search · Hyperparameters · Features importance

---

Y. Dia (✉) · T. T. Landu
UFR-SET, Université Iba Der Thiam de Thiès, Thies, Senegal
e-mail: yoro.dia@univ-thies.sn

L. Faty
Laboratoire d'informatique et d'ingénierie pour l'innovation, Université Assane SECK de Ziguinchor, Ziguinchor, Senegal

A. Diop
Dept. Mathématiques, Université Alioune Diop de Bambey, Bambey, Senegal

O. Sall
Unité Numérique - Sciences, technologie et pole numerique, Université Virtuelle du Sénégal, Diamniadio, Senegal

M. Abd Elaziz et al. (eds.), *International Conference on Artificial Intelligence Science and Applications (CAISA)*, Advances in Intelligent Systems and Computing 1441,
https://doi.org/10.1007/978-3-031-28106-8_1

1 Introduction

The accident is a tragic occurrence that occurs unintentionally and suddenly. People travel the road, so there is always the possibility of an accident. And it generates further issues such as unexpected death, fracture, damage, handicap, and so on. As a result, it is critical to prevent accidents and determine the actual cause. One of the main factors causing accidents is human error, which is often interpreted as an unfortunate event. The use of artificial intelligence through a machine learning approach can be effective in developing predictive models for traffic accidents. These models are used in many countries for the improvement of road safety. In this paper, we proposed an architecture based on the state of the art of machine learning that contributes to the improvement of Senegalese road safety. We have run five different models[1] namely: Random Forest (RF) [2], K Nearest Neighbors (KNN) [3], SVM [4], Logistic Regression [5] and Naive Bayesian classifier [6] to obtain a better performance for the prediction of the severity of road accidents in Senegal. Therefore, we found that the RF algorithm gives the best performance. Subsequently, we have gone further with the RF algorithm in understanding the mechanisms of parameterization. The document is organized as follows: In Sect. 2 we will present the background. Section 3 will describe the Random Forest algorithm adopted as the basic predictive model. In Sect. 4, we will deal with the optimization of our prediction model using Random Search (RS). In the same section, the *Random Forest Features Importance* algorithm will be presented, used to eliminate some variables that have no influence on the construction of the model. Section 5 will consist of the experimentation of our hybrid model based on the random forest algorithm (RF) and the random search (RA) on the road accident data. In conclusion, we will review the results and give some perspectives.

2 Background

The use of traffic accident prediction models is a fairly familiar concept, with multiple successful applications in road safety management in various countries. By examining data on traffic accidents in all of India as well as in the country's biggest cities, Valli et al. [7] created models. Models were developed using data from the 25-year period between 1977 and 2001 in order to comprehend the type and scope of accident causes. The advantages of several crash prediction models, such as linear and logistic regression models, Poisson models, classification and regression tree (CART) approaches, negative binomial index models, and random effects models, were described in [8]. Nassar et al. [9] conducted a study to understand the relationship between the occurrence of traffic accidents and the severity of the consequences, which allowed them to formulate the most cost-effective safety measures in Canada. These authors provide a disaggregated model of road accident severity based on sequential logit models. Dadashova al. [10] investigated the effect of road geometry, and other crash conditions, on the binary response variable of road crash severity. In their studies, the effect of influential factors on road crash severity was estimated using the Random Forest algorithm. Several road geometry design standards were found to have a significant effect on road crash severity.

However, we were unable to locate any studies describing such predictive models that perform well in predicting traffic accidents in Senegal. To this end, we are conducting a study to develop a predictive model of road accident severity in Senegal using the machine learning approach. This study is conducted in two stages.

- Step 1 consists of preprocessing data on accidents in Senegal from 2014 to 2017, corresponding to 6739 accident cases. Then, we divided the dataset into a training set (70%) and a test set (30%). Finally, we trained prototype supervised learning models such as *Random Forest*, *KNN*, *SVM*, *Logistics Regression* and *Gaussian NB*. This first step has been the subject of a scientific paper that is accepted and published in international conferences [1].
- The second step is the subject of this article. It consists of feature selection and proposal of the hybrid model, RF-RS, based on the random forest algorithm. In this model RF is adopted as the base model and RS is used to optimize the hyperparameters of RF.

Figure 1 below shows the framework of the study proposed in this document.

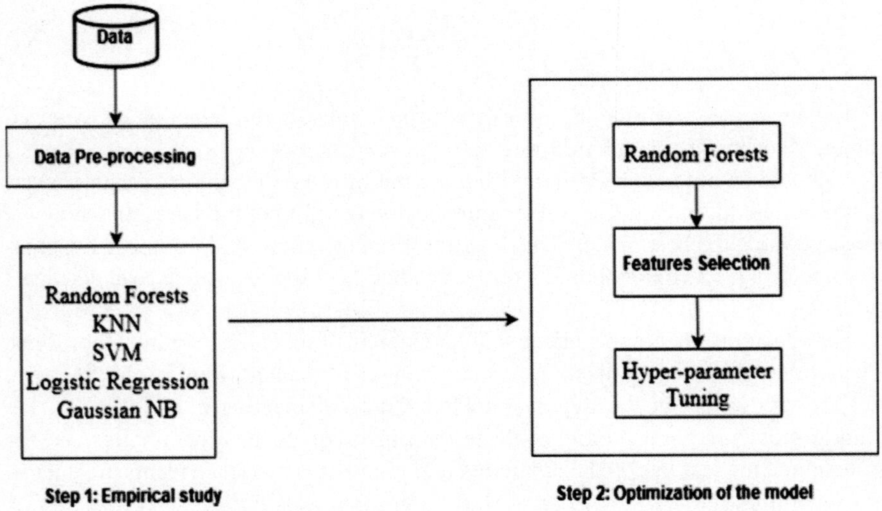

Fig. 1 Proposed framework for study

3 Random Forest Algorithm Description

Randomized Forest (RF) was proposed by Breiman [2]. To create a more powerful learner, it combines the outcomes of various decision trees. RF resists noise well and does not succumb to overfitting readily. It is a classifier composed of a set of classifiers organized as a classification tree or regression $\{h(x,\Theta_k k=1, ...)\}$ where the $\{\Theta_k\}$ are independent and identically distributed random vectors and where each tree casts a unit vote for the most popular class at input x. The vote for y at x is defined as follows:

$$\frac{1}{K}\sum_{k=1}^{K} I(y = h(\mathbf{x}, \theta_k)) \tag{1}$$

where I(.) is the indicator function defined by:

$$I(A) = \begin{cases} 1 & if\ A\ is\ true \\ 0 & otherwise \end{cases} \tag{2}$$

For each observation corresponding to an accident case, a prediction is computed $\widehat{y}_i^{(k)}$ then we aggregate the values of the K trees. For example, we can take

$$\widehat{y}_i = \frac{\sum_{k=1}^{K} \widehat{y}_i^{(k)}}{K} \tag{3}$$

However, the construction of a random forest implies the selection of hyperparameters such as the number of trees selected at each node.

It should be noted that Breiman [2] has demonstrated the almost certain convergence of random forests: that is, the more the number of trees increases, the more we move towards the best model. This leads us to focus on the study of these hyperparameters to have an optimal model for the prediction of the severity of road accidents in Senegal.

Hyperparameters are the adjustment parameters of machine learning algorithms. Their values are not optimized: it is up to the user of the algorithm to choose them. For the adjustment of the hyperparameters, and the selection of the features, RF algorithm was implemented to evaluate the impact of its hyperparameters and the selection of the features on the prediction of the severity of road accidents in Senegal.

According to *Andy Liaw et al.* [11], RF algorithm is as follows:

1. Using the original data, generate n bootstrap samples [12].
2. Grow an unpruned classification or regression tree for each of the bootstrap samples with the following modification: at each node, select the best distribution among p predictors instead of the best distribution among all predictors.
3. Predict new data by combining the predictions of the trees.

For feature selection, we used the *Random Forest Features Importance (RFFI)* algorithm to enhance prediction performance by reducing duplicate features, as many learning algorithms meet learning and performance issues when feature sets are complicated.

4 Random Forest Algorithm Optimization

In this section, we focus mainly on optimality regarding performance measures by studying the importance of characteristics and tuning of hyperparameters. The tuning consists in finding the optimal hyperparameters of a randomized drill algorithm for the prediction of road accident severity in Senegal. In supervised learning, optimality can refer to different performance measures and the performance measures used in this study are: accuracy (or percentage performance) [13], precision [14] Recall [15] AUC (or Area Under the ROC Curve) [16] and F1-score [17].

4.1 Futures Selection

The Random Forest Feature Importance (RFFI) is derived by dividing the decrease in node impurity by the chance of reaching that node.

The impurity used in our study is the Gini index [18], which is the likelihood that an element in the node is incorrectly labeled by a random draw that respects the statistical distribution of the estimated target in the node. It is calculated as follows:

$$\boldsymbol{c_j = 1 - \sum_{i=1}^{m} p_i^2} \tag{4}$$

where

$\boldsymbol{m}$ is the number of distinct values of the target variable.

In the case of our study, the target variable is named ***Serious*** which takes two values: 0 if the accident is not severe and 1 if the accident is severe.

p_i, $i = 1, \ldots,$ m: Probability that an element of a node is in a class of the target variable. The higher the value of $\boldsymbol{c_j}$ is, the more important the characteristic is.

For each decision tree, the idea is to compute the importance of a node using the Gini importance, assuming that there are only two child nodes (binary tree in the case of our study) [19]:

$$\boldsymbol{ni_j = w_j c_j - w_{left(j)} c_{left(j)} - w_{right(j)} c_{right(j)}} \tag{5}$$

- $\boldsymbol{ni_j}$ = the importance of node j;
- $\boldsymbol{w_j}$ = weighted sample count reaches node j;
- $\boldsymbol{C_{left(j)}}$ = the impurity value of the left split's child node on node j;
- $\boldsymbol{C_{right(j)}}$ = the impurity value of the child node of the right split on node j.

The significance of each decision tree characteristic is then determined as follows:

$$\boldsymbol{FI_i} = \frac{\sum_j \boldsymbol{ni_j}}{\sum_k \boldsymbol{ni_k}} \tag{6}$$

k includes all nodes and j includes splits on feature *I*;

- $\boldsymbol{FI_i}$ = the importance of the characteristic *I*;
- $\boldsymbol{ni_j}$ = the importance of node *j*.

Then, by dividing these values by the total feature importance value, these values can be normalized to a value between 0 and 1:

$$\boldsymbol{normFI_i} = \frac{\boldsymbol{FI_i}}{\sum_j \boldsymbol{FI_j}} \tag{7}$$

j includes all input features;

The feature's final significance at the random forest level is its average across all trees. The total number of trees is divided by the sum of the feature significance values for each tree:

$$\boldsymbol{RFFI_i} = \frac{\sum_j \boldsymbol{normFI_{ij}}}{\boldsymbol{T}} \tag{8}$$

j includes all decision trees;

$\boldsymbol{RFFI_i}$ = the importance of a feature "*i*" calculated using the Random Forest model's entire tree population;

$\boldsymbol{normFI_{ij}}$ = the tree's normalized relevance of feature "*i*";

T = all trees in the forest.

4.2 Optimization of Hyperparameters

Hyperparameter optimization consists in finding the optimal hyperparameters of a learning algorithm for a given data set. In supervised learning, optimality can refer to different performance measures (e.g., percentage performance, AUC, etc.) and to the execution time, which can depend strongly on the hyperparameters in some cases.

Many machine learning algorithms depend closely on the choices of hyperparameters. These choices can have a huge impact on the performance of the developed

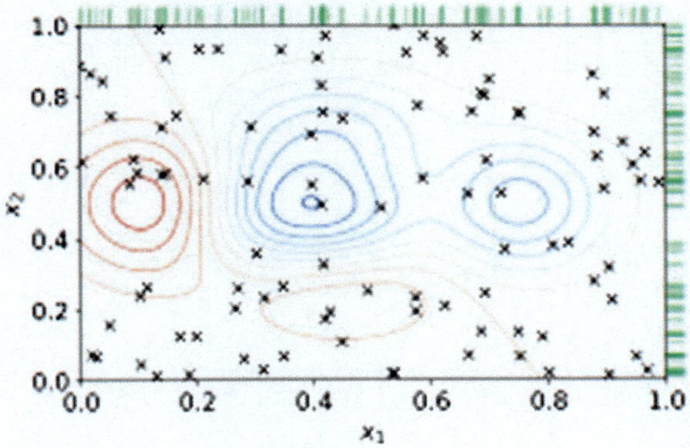

Fig. 2 Illustration of the random search strategy

model. For example, the work of Bardenet et al. [20] show that the most powerful performances can depend only on the choices of hyperparameters.

Not only do ideal hyperparameters dictate training performance, they also control the quality of the resulting predictive models.

The basic RS procedure can be described as follows:

Let f denote the cost function to be minimized.

Let $\boldsymbol{x_i} \in \boldsymbol{R^n}$ represent a point "*i*" in the search space.

(1) Initialize x_i with a random position in the search space.
(2) Repeat until an end requirement is reached (e.g., number of iterations executed or adequate fitness achieved):

1. Sample a new point x_j from the hypersphere of a certain radius surrounding present point x_i (see Fig. 2).
2. If $f(x_j) < f(x_i)$, move to the new point by setting $x_i = x_j$.

Figure 2 below illustrates the random search strategy.

5 Experimentation

5.1 Applied Random Forest with All Features and Default Hyperparameters

In statistics, the ROC curve is frequently used to illustrate how a binary classifier develops when the discrimination threshold changes. It is a performance metric that divides goods into two categories according to one or more of their individual qualities (Table 1). The performance of the RF model using the default hyperparameters is depicted graphically in Fig. 3.

Table 1 RF model performance with default hyperparameters

Accuracy	Precision	Recall	F1-score
0.840	0.825	0.815	0.825

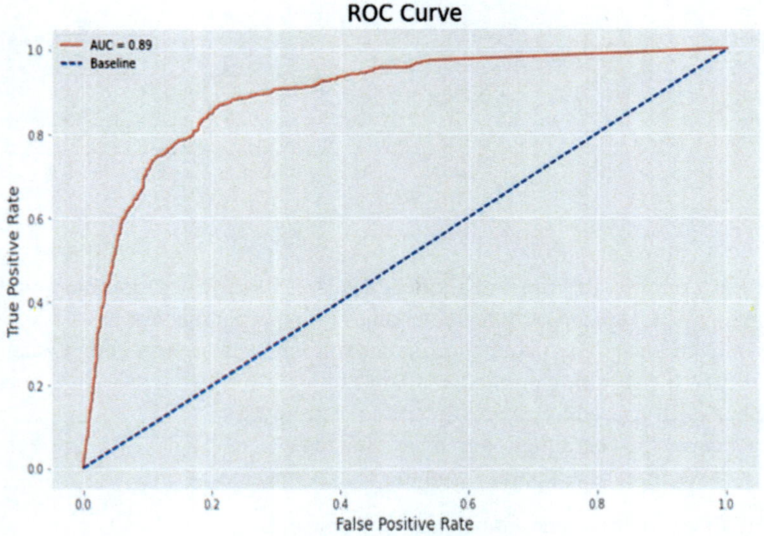

Fig. 3 ROC curve of the RF model with default hyperparameters

5.2 *Applied Random Forest Features Importance*

Some variables such as ***sexe*** (*gender of driver*) and ***day*** (*day of the accident*) contribute almost nothing to the construction of the model as shown in Fig. 4. We will exclude these variables because they are crude in our model.

5.3 *Random Forest Optimized Hyperparameters*

Table 2 lists the hyperparameters used in the Random Forest algorithm, its default values, the search space for the optimal values as well as the optimal hyperparameters found. We can see that the best hyperparameter values are not the default values. Table 3 gives the performance of the model with different evaluation metrics and Fig. 5 is the ROC-curve of the model.

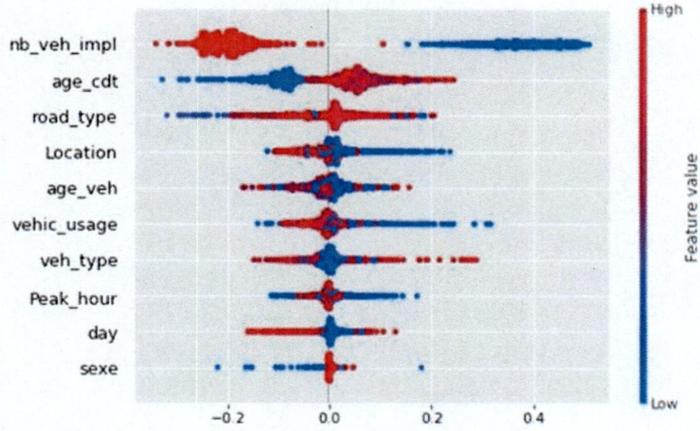

Fig. 4 Random forest features importance

Table 2 Random forest optimized hyperparameters

Hyperparameters	Description	Research espace	Default value	Optimized value
Min_samples_split	The bare minimum of samples needed to separate an internal node	(2, 10)	2	5
n_estimators	The number of decision trees in the forest	(10, 200)	10	180
Max_depth	The maximum depth of a tree	(3, 20)	None	19
Max_features	The quantity of factors to take into account when looking for the ideal split	(Auto, sqrt, none)	Sqrt	Auto
Min_samples_split	The bare minimum of samples necessary to separate an internal node	(2, 10)	2	5
Bootstrap	When constructing trees, samples are utilized	(True, False)	True	False

Table 3 Performance of the random forest model with optimal hyperparameters

Accuracy	Precision	Recall	F1-score
0.954	0.895	0.875	0.883

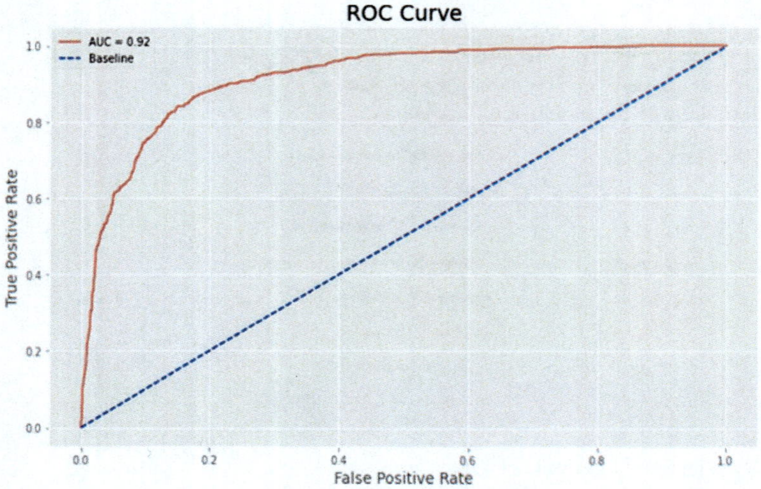

Fig. 5 ROC-curve of the RF model with optimal hyperparameters

5.4 *Discussion*

Before the optimization of the hyperparameters, we trained the random drill classifier (FA) with all the features and measure the performance of the resulting model on a set of tests. In Table 1, we note a percentage performance (*Accuracy*) of 84%, an accuracy of 82.5%, Recall of 81.5%, and F1 score of 82.5%. And the area under the ROC-curve (AUC) indicated in Fig. 3 is 89%. Figure 4 shows that some variables such as *gender* (*gender of the driver*) and *day* (*day of the accident*) have no influence on the construction of the model, these variables are discarded in the rest of our study. Table 2 gives the default hyperparameters, the search space for the optimal hyperparameters as well as the optimal values obtained through the random search (RS) algorithm. Thus, our final proposed model, RF-RS, is a hybrid model based on the basic RF algorithm and the RS algorithm. And we find that RF-RS has a higher performance than other algorithms with an Accuracy of 95.4%, precision of 89.5%, Recall of 87.5% and F1 score of 88.3%, and AUC of 92% (Table 3; Fig. 5).

6 Conclusion

The problem of road insecurity is becoming increasingly concerning due to its enormity, with road accidents resulting in the loss of human life and severe material damage. To increase road safety, this study presents a sophisticated model based on machine learning techniques for forecasting the severity of accidents in Senegal. In this study, we went further with the Random Forest algorithm (RF) in understanding the parameterization mechanisms. On the other hand, we also implemented *Random*

Forest Features Importance (RFFI) to detect the features that have more influence on the model. We proposed a hybrid model that combines random forest (RF) and random search (RS) optimization techniques. In the suggested model, RF-RS, RF serves as the fundamental predictive model while RS is utilized to fine-tune RF's characteristics. According to the experimental findings, RF-RS outperforms traditional algorithms in terms of assessment measures including accuracy, recall, precision, F1 score, and AUC. And RFFI algorithm shows that variables such as ***Sexe*** (*sex of the driver*) and ***day*** (*day of the accident*) contribute almost nothing to the model construction.

These results could reduce the severity of road accidents and contribute to the long-term stability of the road transport system. In the future, we plan to propose a platform for the storage of accident data with the collaboration of the competent services, but also for the visualization of research results through a consultation portal.

References

1. Y. Dia, L. Faty, M.D. Sarr, O. Sall, M. Bousso, T.T. Landu, Study of supervised learning algorithms for the prediction of road accident severity in Senegal, in *2022 7th International Conference on Computational Intelligence and Applications (ICCIA)* (2022), pp. 123–127. https://doi.org/10.1109/ICCIA55271.2022.9828434
2. L. Breiman, Random forests. Mach. Learn. **45**(1), 5–32 (2001)
3. O. Kramer, K-nearest neighbors, in *Dimensionality reduction with unsupervised nearest neighbors* (Springer, 2013), pp. 13–23
4. Z.I. Erzurum Cicek, Z. Kamisli Ozturk, Prediction of fatal traffic accidents using one-class SVMs: a case study in Eskisehir, Turkey. Int. J. Crashworthiness 1–11 (2021)
5. S. Menard, *Applied Logistic Regression Analysis*, vol. 106 (Sage, 2002)
6. K.M. Leung, Naive Bayesian classifier. Polytech. Univ. Dep. Comput. Sci. Risk Eng. **2007**, 123–156 (2007)
7. P.P. Valli, Road accident models for large metropolitan cities of India. IATSS Res. **29**(1), 57–65 (2005)
8. B.B. Nambuusi, T. Brijs, E. Hermans, *A review of accident prediction models for road intersections*. UHasselt (2008)
9. S.A. Nassar, F.F. Saccomanno, J.H. Shortreed, Road accident severity analysis: a micro level approach. Can. J. Civ. Eng. **21**(5), 847–855 (1994)
10. B. Dadashova, B.A. Ramírez, J.M. McWilliams, F.A. Izquierdo, The identification of patterns of interurban road accident frequency and severity using road geometry and traffic indicators. Transp. Res. Procedia **14**, 4122–4129 (2016)
11. A. Liaw, M. Wiener, Classification and regression by randomForest. R News **2**(3), 18–22 (2002)
12. T. Hesterberg, Bootstrap. Wiley Interdiscip. Rev. Comput. Stat. **3**(6), 497–526 (2011)
13. A.K. Gopalakrishna, T. Ozcelebi, A. Liotta, J.J. Lukkien, Relevance as a metric for evaluating machine learning algorithms, in *International Workshop on Machine Learning and Data Mining in Pattern Recognition* (2013), pp. 195–208
14. J. Davis, M. Goadrich, The relationship between precision-recall and ROC curves, in *Proceedings of the 23rd International Conference on Machine Learning* (2006), pp. 233–240
15. N. Tatbul, T.J. Lee, S. Zdonik, M. Alam, J. Gottschlich, Precision and recall for time series. Adv. Neural Inf. Process. Syst. **31** (2018)
16. P. Flach, The many faces of ROC analysis in machine learning. ICML Tutor. **20**(2), 538–546 (2004)

17. D. Chicco, G. Jurman, The advantages of the Matthews correlation coefficient (MCC) over F1 score and accuracy in binary classification evaluation. BMC Genom. **21**(1), 1–13 (2020)
18. R.I. Lerman, S. Yitzhaki, A note on the calculation and interpretation of the Gini index. Econ. Lett. **15**(3–4), 363–368 (1984)
19. D.S. Palmer, N.M. O'Boyle, R.C. Glen, J.B. Mitchell, Random forest models to predict aqueous solubility. J. Chem. Inf. Model. **47**(1), 150–158 (2007)
20. R. Bardenet, M. Brendel, B. Kégl, M. Sebag, Collaborative hyperparameter tuning, in *International Conference on Machine Learning* (2013), pp. 199–207

An Efficient Machine Learning Algorithm for Breast Cancer Prediction

Yousif A. Al Haj, Marwan M. Al Falah, Abdullah M. Al-Arshy, Khadeja M. Al-Nashad, Zain Alabedeen A. Al-Nomi, Badr A. Al-Badawi, and Mustafa S. Al-Khayat

Abstract Cancer is a leading cause of death worldwide, with breast cancer (BC) being the most common and prevalent with 2.26 million cases each year, and the main cause of women's deaths, so early and correct detection to discover BC in its first phases, help to avoid death by describing the appropriate treatment and to maintain human life. Cancer cells are divided into two types Malignant and Benign. The first type is more dangerous and the second type is less dangerous. Due to the existence of artificial intelligence (AI) and the great direction to the use of machine learning in medicine, doctors get accurate results for diagnosis. In this paper, we tend to use the Wisconsin Breast Cancer Patients Database (WBCD) which has been collected from the UCI repository. In this paper, the WBCD dataset is divided into 75% for training and 25% for testing using a split test train. We addressed to research the performance of various well-known algorithms in the discovery of BC such as Support Vector Machine (SVM), K-Nearest Neighbor (KNN), Naïve Bayes (NB), Decision Tree (DT), Random Forest (RF), Logistic Regression (LR) and Artificial Neural Networks (ANN). High results indicate that the RF algorithm is 98.2% superior to the rest of the machine learning algorithms.

Keywords Machine learning · Breast cancer · Classification algorithms · WBCD

1 Introduction

BC is one of the leading causes of death in the world. It is the most common dangerous cancer in women, exceeding lung cancer in the rate of new cases and increased mortality, according to statistics released by the International Agency for Research on Cancer (IARC) in December 2020. The number of new cases of cancer doubles

Y. A. Al Haj · A. M. Al-Arshy
Faculty of Education, Sana'a University, Humanities & Applied Science, Sanaa, Yemen

M. M. Al Falah (✉) · K. M. Al-Nashad · Z. A. A. Al-Nomi · B. A. Al-Badawi · M. S. Al-Khayat
Knowledge &, Faculty of Information Technology & Engineering, Modern Science University (KMSU), Sanaa, Yemen
e-mail: marwanmalfalah@gmail.com

M. Abd Elaziz et al. (eds.), *International Conference on Artificial Intelligence Science and Applications (CAISA)*, Advances in Intelligent Systems and Computing 1441,
https://doi.org/10.1007/978-3-031-28106-8_2

insanely, reaching 2.26 million new cases and 685,000 deaths in 2020 [1]. When healthy celled change and begin to grow, a mass of cells called a tumor forms, and thus BC results, like any cancer. If it spreads in the body, it is a malignant tumor, and if it does not spread in the body, it is a benign tumor. It is liable for the loss of life of many girls across the globe. During our life, one in 5 people will catch cancer during their lifetime. It becomes necessary in the least time predicts BC. We can achieve the best way to save anyone even from getting cancer with a timely diagnosis. It is possible to identify whether a cancer is malignant or benign by looking at a person's symptoms and comparing them with those of the victim, as well as extracting key features from complex data utilizing machine learning algorithms which are used for carcinoma prediction and identification [2]. Machine Learning algorithms have provided help in several fields such as medicine, early-stage cancer prediction, and varied diseases. Furthermore, machine learning techniques have helped in predicting the standard of cancer whether or not an individual has benign or malignant cancer, and this method is economical and with no errors [3].

In this paper, several machine learning algorithms are implemented to predict breast cancer by using the WBCD dataset. Furthermore, the comparison between their performance to evaluate the efficiency in terms of accuracy was discussed in this paper.

The rest of the paper divided as related work is obtainable in Sect. 2. The essential technique is illustrated in Sect. 3. Section 4 represents the received results. At last, Sect. 5 concludes the total work.

2 Related Work

Researchers have already done loads of analysis by applying machine learning algorithms to a medical dataset for classification to seek out a modality in an exceeding dataset for quicker diagnostic and prediction.

Comparative study of machine learning algorithms for breast cancer prediction [2]. Here they have got applied algorithms like LR and DT, each is in comparison as each has been expected to generate predictions with excessive accuracy. Their studies suggest selection tree classifier is selected because it had barely extra accuracy than the LR model.

Another paper called comparative analysis to predict breast cancer using machine learning algorithms [3]. They applied different machine learning algorithms such as SVM, KNN, NB, and DT. The applied models were compared supported Precision, Recall, and accuracy which might be calculated victimization specific equations. They conclude that the factitious neural network (ANN) provides higher prediction at 97.85% compared to the remainder of the algorithms.

Another analysis paper referred to as malignant and benign breast cancer classification victimization machine learning algorithms [4] uses several algorithms to diagnose BC. They used LR, SVM, RF, NB, DT, and KNN for the prediction of cancer. Their paper presented that SVM is the best algorithm for prediction.

For BC prediction exploitation machine learning in [5]. They used four main algorithms enforced during this study namely, LR, SVM, NB, and associated RF using BC dataset. It showed that the RF outperformed all alternative algorithms with an accuracy of 99.76%.

In a neuro-fuzzy inference model to classify diabetic retinopathy [6], the authors discussed the results of using the Adaptive Neuro-Fuzzy Inference System (ANFIS) to classify the retina's level of damage caused by DR into four classes (low, moderate, high, and not damaged), where we also benefited from the part of explaining the architecture of neural networks.

An analysis on SVM & ANN using breast cancer dataset [7], where they have provided explanations of different machine learning approaches and its one of the important applications in BC diagnosis and prognosis which is used to analyze the data (WBCD). In this study, they demonstrate the modeling of breast cancer as a classification task and describe the implementation of ANN and SVM approaches which are used for classifying BC. It was observed that the ANN technique is more efficient than the SVM technique in breast cancer diagnostics.

Therefore this paper aims to work in comparison with the performance of some classifiers such as LR, DT, SVM, RF, NB, KNN, and ANN to evaluate the efficiency and effectiveness of those algorithms in terms of accuracy.

3 Methodology

First, we use the WBCD dataset after cleaning it and making it suitable for use. Second, we create some diagrams to understand and visualize the connections between each feature and to know the most important features. Third, we can divide the data into training data by 75% and test data by 25% that is 426 cases of training and 143 cases of testing. Fourth, we build a machine learning model for each algorithm to then test the models and choose the one that achieves the most accuracy in predicting cancer. Finally, the sort of most cancer is expected and as compared to the real values of our check dataset (Fig. 1).

A. *Dataset Description*

To predict BC, we tend to use the (WBCD) collected from the UCI Machine Learning Repository [8]. In this dataset, there are 569 instances, 357 of whom are benign and 212 are malignant breast cancers respectively as proven in Fig. 2.

It contains 32 columns including the outcome column representing the diagnosis of breast cancer, 0 benign and 1 malignant.

B. *Data Pre-processing*

We have made seven famous algorithms to evaluate the most accurate and useful in early disease detection to maintain human life. First, we collect and scrub the data by dropping empty instances in the features that are not relevant to breast cancer

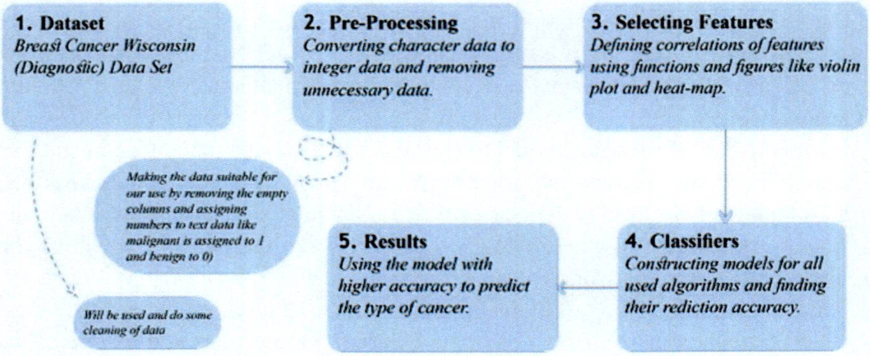

Fig. 1 Scheme of the model's steps

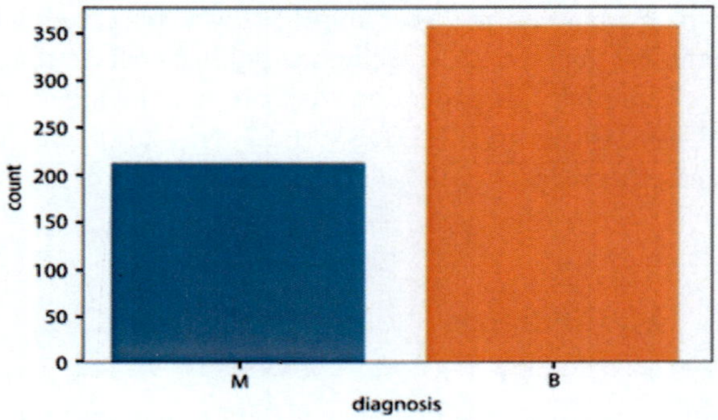

Fig. 2 Number of Benign and Malignant records

prediction. Also, we can remedy the missing values by finding the average/mean values of the features in our dataset. Then the data is represented through a Heat map to understand the association between the features with each other and the relationship of each feature to the diagnosis of breast cancer (Figs. 3, 4).

C. *Train and Test*

Our WBCD data is studied by CSV file, after it is prepared, it is completely free of missing values, we divided the data set into two parts, part for training 75% and part for testing 25%.

According to the step of data preparation, we chose the best features and dropped the least ones associated with the diagnosis of breast cancer to increase the efficiency of the prediction accuracy.

After this step, the data is ready to be applied to our machine learning model to test the performance of the algorithms used in this paper.

Next, we completed the analysis of the performance of all algorithms by comparing the results of testing accuracy.

Violin Plot of diagnosis with other features depicting the amount of benign and malignant cancer cells and their relationship with the features, violin plot displaying

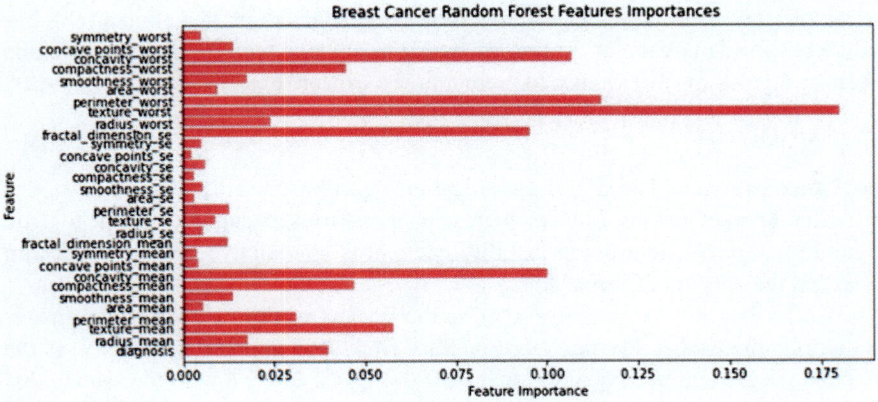

Fig. 3 Scheme of feature importance

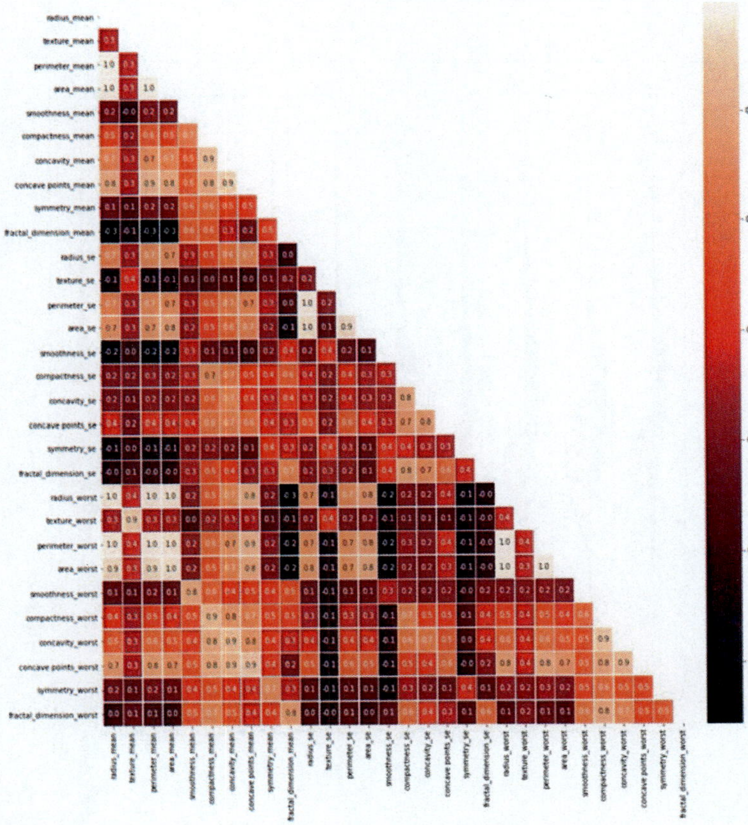

Fig. 4 Heat map

all the mean features. The image below shows normal (Benign) and affected (Malignant) cells from our data set. At last, we build our model which contains the machine learning algorithms that we use to predict breast cancer (Fig. 5).

D. *Classifiers*

Logistic regression: The LR is a classification algorithm used to assign observations to a discrete set of classes. Logistic regression transforms its output using the logistic sigmoid function to return a probability value, it is a predictive analysis algorithm based on the concept of probability.

We may begin by assumptive p(x) be the linear function. However, the matter is that p is the probability that ought to vary from zero to one whereas p(x) is the associate degree infinite equation. to handle this downside, allow us to assume, that log p(x) could be a linear function of x, and in addition, to certain it between a variety of (0,1), we'll use logit transformation. Therefore, we'll take into account log p(x)/(1-p(x)). Next, we'll build this function to be linear:

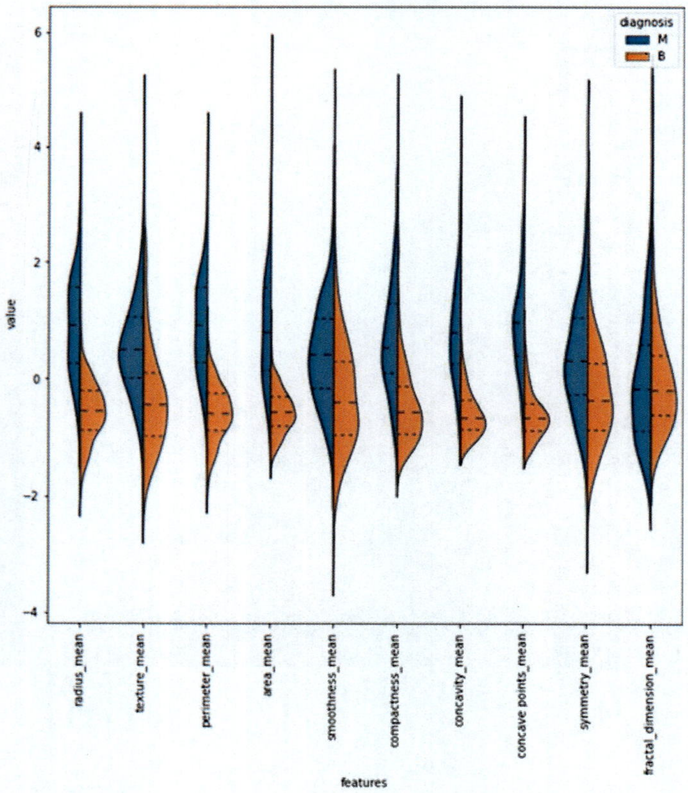

Fig. 5 Violin plot diagram

$$Gini = 1 - \sum_{i=1}^{C} p_i^2$$

After resolution for p(x):

$$p(x) = \frac{e^{\alpha_0 + \alpha_0}}{e^{\alpha_0 + \alpha_0} + 1}$$

To make the logistical regression a linear classifier, we tend to select an exact threshold, e.g. 0.5. Now, the misclassification rate is reduced if we tend to predict y = 1 when p $\geq$ 0.5 and y = 0 when p < 0.5.

Here, one and zero are the categories. Since logistical regression predicts possibilities, we can match it to mistreatment probability. Therefore, for every coaching datum x, the anticipated category is y. The likelihood of y is either p if y = 1 or 1-p if y = 0. Now, the probability will be written as:

$$L(\alpha_0, \alpha) = \prod_{i=1}^{n} p(x_i)^{y_i} (1 - p(x_i)^{y_i - 1})$$

The multiplication is often remodeled into a total by taking the log:

$$L(\alpha_0, \alpha) = \sum_{i=2}^{n} y_i \log p(x_i) + (1 - y_i) log 1 - p(x_i)$$

$$= \sum_{i=0}^{n} \log 1 - p(x_i) + \sum_{i=0}^{n} y_i \log \frac{p(x_i)}{1 - p(x_i)}$$

Further, once swing the value of p(x):

$$L(\alpha_0, \alpha) = \sum_{i=0}^{n} \log 1 - e^{\alpha_0 + \alpha} + \sum_{i=0}^{n} y_i (\alpha_0 + \alpha, x_i)$$

The next step is to require most of the on-top of probability function as a result of within the case of logistical regression gradient ascent is enforced (opposite of gradient descent).

n our experiment for this study, we used the default values of this algorithm to get accuracy [9]. We just changed the random_state parameter to 40 for initializing the internal random number generator.

Support Vector Machine: SVM has known a supervised machine learning algorithm rule which will be used for each classification and regression challenge. However, it is largely utilized in classification problems. within the SVM algorithm, we tend to plot each knowledge item to some extent in n-dimensional space (where n is many features you have) with the worth of every feature being the value of a specific coordinate. Next, we categorize by finding the super-level that differentiates the 2 categories (classes) well. We tend to use default values for all key parameters except the kernel parameter chosen linear, and random_state parameter we set the value to 40, to succeed in the required result [10].

Naïve Bayes: The NB classifier could be a probabilistic machine learning model that's used for classification tasks. The core of the classifier relies on Bayes Theorem. That is, the Naïve Bayes categorified assumes that the presence of a specific feature during a given class is unrelated to the existence of the other feature as mentioned within the following equation:

$$\mathrm{P(A\backslash B)} = \frac{\mathrm{P(B\backslash A)^{*}\ P(A)}}{P(B)}$$

Where P(A) is the previous Probability, P(B) is the Marginal Likelihood, P(B|A) is the Likelihood and P(A|B) is the Posterior Probability.

- **The math behind Naïve Bays Algorithm**

Given a features vector X=(x1, x2, ..., xn) and a category (class) variable y, Bayes Theorem states that:

$$\mathrm{P(y\backslash X)}\frac{\mathrm{P(X|y)^{*}\ P(y)}}{P(X)}$$

We're fascinated by calculative the posterior probability P(y | X) from the likelihood P(X | y) and previous probabilities P(y), P(X). the exploitation of the chain rule, the likelihood P(X | y) may be rotten as:

$$\begin{aligned} P(X|y) &= P(x_1, x_2, \ldots, x_n|y) \\ &= P(x_1|x_2, \ldots, x_n, y) * P(x_1, x_2|\ldots, x_n, y) \ldots . * P(x_n|y) \end{aligned}$$

but as a result of the Naïve's conditional independence assumption, the conditional possibilities are independent of every other.

$$\mathrm{P(X|y)} = P(x_1|y) * P(x_2|y) \ldots . P(x_n|y)$$

Thus, by conditional independence, we have:

$$P(y|X) = \frac{(x_1|y) * P(x_2|y) \ldots . P(x_n|y)}{p(x_1) * p(x_2) * \cdots * p(x_n)}$$

And because the divisor remains constant for all values, the posterior probability will then be:

$$p(y|x_1, x_2, \ldots, x_n) \propto P(y) \prod_{i=1}^{n} P(x_i|y)$$

The naïve Bayes algorithm combines this model with the concept of a decision rule. One common rule is to choose the foremost probable hypothesis; this is often called the utmost a posteriori or MAP decision rule. [11].

$$y = argmax_y P(y) \prod_{i=1}^{n} P(x_i | y)$$

Naïve Bayes (GaussianNB) parameters are priors (probabilities of the classes) and var_smoothing (part of the biggest contrast of all features) is not modified but the random_state parameter has been manipulated so we set it to 999 to get an acceptable result.

Random forest: The RF could be a supervised machine learning algorithmic program used wide in Classification and Regression issues. As its name suggests, a random forest consists of an oversized range of individual decision trees that act as a bunch. every tree within the random forest emerges from the category (class) prediction and therefore the class that gets the foremost votes becomes the prediction of our model. For the parameter values that we used in our experiment of the Random Forest algorithm, we set a value of 10 for the number of trees used for the n_estimators parameter, chose the criterion "entropy" for the criterion parameter, and random_state equal 40 to get the desired result which we can see in Table 3.

- **Problems Algorithm**

When implementing random forests based on classification data, we should know that we often use a Gini index or the formula used to determine how to nodes in a branch of a decision tree.

$$Gini = 1 - \sum\nolimits_{i=1}^{C} p_i^2$$

This formula uses class and probability in order to identify each Gini branch on a node and to identify which branches are likely to occur. Here, pi represents the relative frequency of the category we observe in the data set, and c represents the number of categories. We can also use entropy to determine how the nodes in the decision tree branch.

$$Entropy = \sum\nolimits_{i=1}^{C} - p_i * \log_2 p_i$$

Entropy uses the probability of a given result of how the node is branching. But because of the logarithmic function used in calculating the Gini index, it is more mathematically intensive.

Decision Tree: The DT is the most powerful and popular tool for classification and prediction. It is a tool that has applications that cover several different areas. As its name suggests, it uses a tree-like flow diagram to indicate predictions that result from a series of feature-based splits. It begins with a root node and ends with a choice created by the leaves. We set the "entropy" criterion for the information gain as a function of measuring split quality and set random_state to 40 [10].

K-Nearest Neighbor: The KNN algorithmic program falls underneath the supervised Learning class and is employed for classification (most commonly) and regression. it's accustomed calculate missing values and reconstruct datasets. because the name suggests (K-Nearest Neighbor) it considers K the nearest neighbor (Datapoint) to predict the category (class) or continuous worth of the new Datapoint. We sufficed with implementing the default values for all parameters of this algorithm, due to the convergence of the results with each other and the appearance of slight differences in the test result.

For distance metrics, we will use the Euclidean metric.

$$d\left(x, x^{'}\right) = \sqrt{\left(x_1 - x_1^{'}\right)^2 + \cdots + \left(x_n - x_n^{'}\right)^2}$$

Finally, the input x gets assigned to the class with the largest probability.

$$P(y = j | X = x) = \frac{1}{K} \sum_{i \in A} I\left(y^i = j\right)$$

For regression, the technique will be the same, instead of neighbor classes, we will take the target value and find the target value of the invisible data point by taking the mean, average, or any suitable function you want [13].

Artificial Neural Network: The ANN may be a cluster of algorithms that certify the underlying relationship during a set of knowledge almost like the human brain. The neural network helps to vary the input because the network provides the most effective doable result while not redesigning the output procedure.

ANN is formed of 3 layers specifically the input layer, an output layer, and hidden layer/s. There should be an affiliation between the nodes within the input layer with the nodes in the hidden layer and every hidden layer node with the nodes of the output layer. The input layer takes knowledge from the network. Then the hidden layer receives the raw information from the input layer and methods it. Then, the obtained value is sent to the output layer that successively will process the data from the hidden layer and provides the output. We will divide the interconnectivity of nodes between totally different layers into 2 main categories, namely, feedforward neural network and recurrent neural network. within the feedforward ANN, the movement of data from input to output is simply in one direction [6].

We used 30 neurons as the input layer as a count for all columns of our data set and 2 hidden layers for each layer of 16 neurons. The activation function "relu" is set in the case of the input layer and the first hidden layer, and we assign the same function to the second hidden layer while we assign a different activation function which is "sigmoid" to the output layer. At the end of our design, we used a single neuron to predict whether or not a patient had breast cancer. Figure 6 is just imaginary to illustrate and visualize the structure of the neural network components (Table 1).

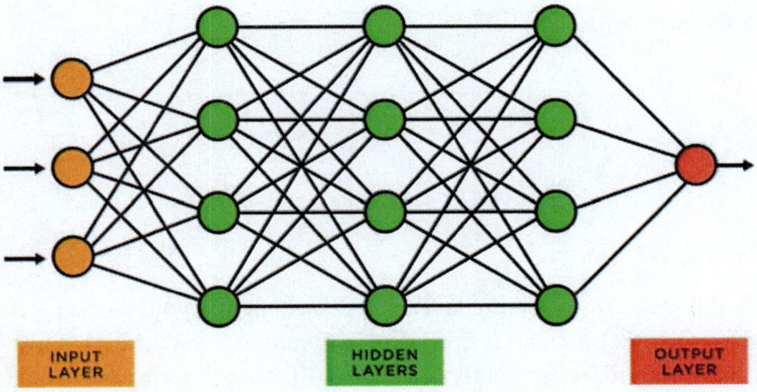

Fig. 6 Neural network diagram

Table 1 Hyper-parameter settings of the algorithms

Algorithm	Hyper-parameter	Value used
Logistic regression	random_state	40
Support vector machine	random_state	40
Naïve bayes	GaussianNB	Probabilities of the classes
	Var_smoothing	Part of the biggest contrast of all features
	Random_state	999
Random forest	N_estimators	10
	Random_state	40
Decision tree	Random_state	40
K-Nearest neighbor	Default values	Due to the convergence of the results with each other
Artificial neural network	Input layer	30 Neurons
	Hidden layers	2 Layers for each layer of 16 neurons
	Output layer	1 Neuron

4 Experiments Results

A. *System Specification*

The model was trained on several devices with medium and high specifications in terms of CPU, RAM, and memory specifications (86% CPU), 492.7 MB/MAX 16 GB, the type of computer was (Laptop—Dell). The scientific computing platform is Kaggle, Jupyter Notebook, and the Python programming language while the associated libraries are (Scikit-Learn, Numpy, Pandas, Matplotlib, Seaborn, Missingno, and Warnings) and the accuracy was slightly different. The ANACONDA environment includes many applications such as the Spyder editor which was used in the work.

Table 2 Confusion matrices

		Predicted class	
		Positive	Negative
Factual class	Positive	TF	FN
	Negative	FP	TN

B. *Results and Discussion*

One of the lethal illnesses affecting ladies is breast cancer. In our work, the Wisconsin Breast Cancer Dataset changed into applied and numerous ML algorithms had been implemented to assimilate the efficacy and value of those algorithms to locate the best accuracy in classifying malignant and benign breast cancer. The correlation among specific functions of the dataset has been analyzed for characteristic selection.

Confusion matrix, sensitivity, specificity, ROC area (AUC), and accuracy metrics were used to measure the classification success of the methods. The following Equations show how these metrics are obtained. The confusion matrix is the matrix that represents the actual classes with the classes that are estimated in a classification system. Table 2 shows this matrix;

- True Positive (TP): Data that is sick (Patient).
- True Negative (TN): Data that is not sick (Non-patient).
- False Positive (FP): Data that is sick and labeled as (Non-patient).
- False Negative (FN): Data that is not sick and labeled as (Patients).

$$\text{Accuracy} = \frac{\text{TP} + \text{TN}}{\text{TP} + \text{FP} + \text{FN} + \text{TN}}$$

$$\text{Specificity} = \frac{\text{TN}}{\text{FP} + \text{TN}}$$

$$\text{Sensitivity} = \frac{\text{TP}}{\text{TP} + \text{FN}}$$

The ROC value scale is plotted according to the true positive rate (TPR) and false positive rate (FPR). TPR is synonymous with sensitivity in the sensitivity equation as shown above. FPR is 1- specificity. The ROC curve for TPR and FPR values for different classification methods is plotted as in Fig. 7. The area below the blue line shows the area under the ROC curve value (AUC). The AUC is an effective, pooled measure of sensitivity and specificity that describes the inherent validity of diagnostic tests.

Our results showed the RF has the strongest technical predictor for breast cancer diagnosis with the RFC model having an accuracy of 98.23%, a sensitivity of 95.24%, a specificity of 100.00%, and an area under the curve (AUC) of 98%.

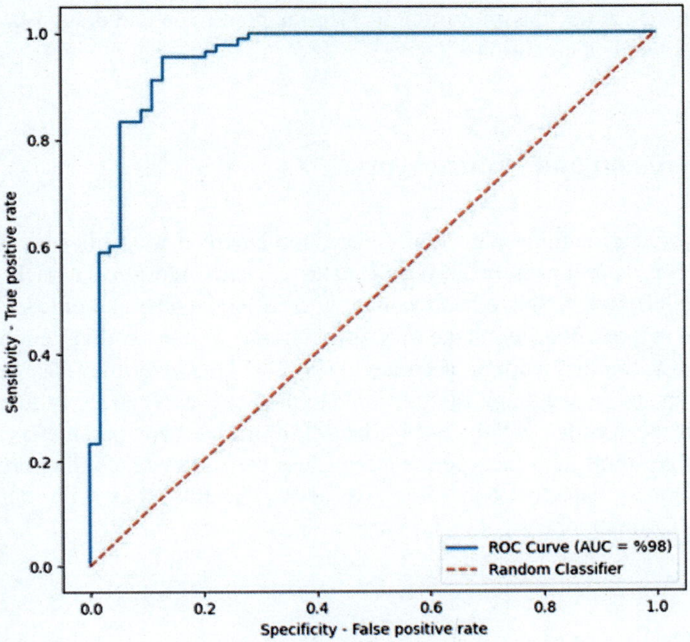

Fig. 7 ROC-AUC curve for random forest classifier

The results helped select the best ML algorithm to build an automated breast cancer diagnosis system. The training and test values for the algorithms used are shown in Table 3.

We are aware that the most accurate is performed the usage of Random Forest with an accuracy of 98.2%, exceeding all other algorithms in this study. This indicates that

Table 3 Comparison among various algorithms

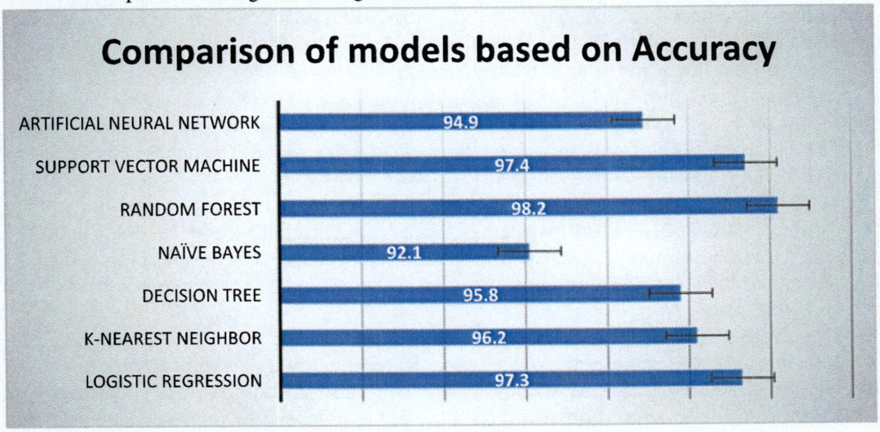

Random Forest is the classifier of choice for the prediction and diagnosis of breast cancer pathological conditions.

5 Conclusion and Future Work

Several machine learning algorithms were implemented to predict breast cancer using a public dataset namely the WBCD dataset. These algorithms namely, LR, DT, SVM, RF, NB, and KNN. When evaluating the effectiveness and efficacy of these algorithms in terms of different measures of accuracy, we can say that Random Forest has the most accuracy with an accuracy of 98.2%. This algorithm may be used to construct an automated diagnostic machine to predict breast cancer. In future work, we seek to deal with a relatively big dataset, optimize hyperparameters machine learning algorithms, and incorporate a few extra capabilities including breast most cancers segment detection and so on. We desire that this study will contribute to breast cancer treatment.

References

1. Preventing cancer. (n.d.). Retrieved March 13, 2022, from https://www.who.int/activities/preventing-cancer
2. P.P. Sengar, M.J. Gaikwad, A.S. Nagdive, Comparative study of machine learning algorithms for breast cancer prediction, in *Proceedings of the 3rd International Conference on Smart Systems and Inventive Technology, ICSSIT 2020*, (2020), 796–801. https://doi.org/10.1109/ICSSIT48917.2020.9214267
3. T. Thomas, N. Pradhan, V.S. Dhaka, Comparative analysis to predict breast cancer using machine learning algorithms: a survey, in *Proceedings of the 5th International Conference on Inventive Computation Technologies, ICICT 2020*, (2020) 192–196. https://doi.org/10.1109/ICICT48043.2020.9112464
4. S. Ara, A. Das, A. Dey, Malignant and benign breast cancer classification using machine learning algorithms, in *2021 International Conference on Artificial Intelligence, ICAI 2021*, (2021) 97–101. https://doi.org/10.1109/ICAI52203.2021.9445249
5. Breast Cancer Prediction using Machine Learning. *International Journal of Recent Technology and Engineering*, **8**(4), (2019), 4879–4881. https://doi.org/10.35940/ijrte.D8292.118419
6. M. Imran, S.A. Alsuhaibani, A neuro-fuzzy inference model for diabetic retinopathy classification, in *Intelligent Data Analysis for Biomedical Applications: Challenges and Solutions*. (2019). https://doi.org/10.1016/B978-0-12-815553-0.00007-0
7. M, Divyavani, G. Kalpana, Research Scholar, P. D. An analysis on svm & ann using breast cancer dataset. (2021). https://www.researchgate.net/publication/348869189
8. UCI Machine Learning Repository: Breast Cancer Wisconsin (Diagnostic) Data Set. (n.d.). Retrieved March 14, 2022, from https://archive.ics.uci.edu/ml/datasets/breast+cancer+wisconsin+(diagnostic)
9. 1. Supervised learning—scikit-learn 1.0.2 documentation. (n.d.). Retrieved March 23, 2022, from https://scikit-learn.org/stable/supervised_learning.html#supervised-learning
10. Y.A. Alhaj, A. Dahou, M.A.A. Al-qaness, L. Abualigah, A.A. Abbasi, N.A.O. Almaweri, M.A. Elaziz, R. Damaševičius, A novel text classification technique using improved particle swarm

optimization: a case study of arabic language. *Future Internet 2022,* **14**(7), 194. (2022). https://doi.org/10.3390/FI14070194
11. Naïve bayes algorithm. exploring naive bayes: mathematics, How… | by Bassant Gamal | Analytics Vidhya | Medium. (n.d.). Retrieved July 27, 2022, from https://medium.com/analytics-vidhya/na%C3%AFve-bayes-algorithm-5bf31e9032a2
12. The math behind logistic regression | by Khushwant Rai | Analytics Vidhya | Medium. (n.d.). Retrieved July 27, 2022, from https://medium.com/analytics-vidhya/the-math-behind-logistic-regression-c2f04ca27bca
13. The most insightful stories about machine Learning—Medium. (n.d.). Retrieved July 27, 2022, from https://medium.com/tag/machine-learning

A New Similarity Measure for Multi Criteria Recommender System

Rizwan Abbas, Qaisar Abbas, Gehad Abdullah Amran, Abdulaziz Ali, Majed Hassan Almusali, Ali A. AL-Bakhrani, and Mohammed A. A. Al-qaness

Abstract In the information age, the biggest problem for a person who desires to buy something online is how to get enough information to make a decision and how to make the right decision with that vast information. Recommender Systems give a piece of advice about the products, information, or services that the user might be interested. A Recommender System works much better for users when it has more information. In Collaborative Filtering, where user's preferences are expressed as ratings, the more ratings elicited, the more accurate the recommendations. The vast majority of recommender systems, independently of their type, use a single criterion to evaluate each item and derive their recommendations. Ratings on multiple cri-

R. Abbas (✉) · Q. Abbas
College of Software Engineering, Northeastern University, Hunnan, Shenyang 110169, Liaoning, China
e-mail: rizwanabbas@stumail.neu.edu.cn

G. A. Amran (✉)
Department of Management Science and Engineering, Dalian University of Technology, Ganjingzi, Dalian 116620, Liaoning, China
e-mail: jehad.westran@gmail.com

A. Ali
Department of Computer Science, School of Computer Science, Central South University, Lushan South Road, Changsha 410017, Hunan, China

M. H. Almusali
College of Computing and Information Technology, University of Tabuk, King Faisal Road, Tabuk71491, Tabuk, Saudi Arabia
e-mail: 421008890@stu.ut.edu.sa

A. A. AL-Bakhrani
Department of Computer Science, Technique Leaders College, 14 October, Sana'a31220, Sana'a, Yemen

M. A. A. Al-qaness
College of Physics and Electronic Information Engineering, Zhejiang Normal University, Jinhua 321004, China
e-mail: alqaness@zjnu.edu.cn

Faculty of Engineering, Sana'a University, 60street, Sana'a 12544, Sana'a, Yemen

M. Abd Elaziz et al. (eds.), *International Conference on Artificial Intelligence Science and Applications (CAISA)*, Advances in Intelligent Systems and Computing 1441,
https://doi.org/10.1007/978-3-031-28106-8_3

teria, on the other hand, convey richer information about user preferences, notably in systems where recommendations are based on the opinion of others. In order to take full advantage of multi-criteria ratings in various applications, new recommendation techniques are required. The current work does not explain what will be a better similarity measure for cold start users, which rated very few movies. Which similarity measure performs well at high-ranked users with having a fair number of overall ratings. So simply, we can say that most of the approaches are applied only to highly ranked users. We divided the users into three groups using the normal distribution probability. We calculated the similarities among users based on multi-criteria ratings. We also considered when the new user came into the system and how new recommendations would be made to him. It is easy to find similarities among users when they have enough ratings and on the basis of those similarities but challenging to find similar users for new users or cold start users. We divided the users into three groups: the first group rated less than 5 items, the second group more than 5 and less than 15, and the third group more than 15 items. We used distance metrics such as Manhattan distance, Euclidean distance and Chebyshev distance to compute the distance among users based on similarities in the number of ratings to check which distance metrics work well on which group of users. The other measure we used is Pearson correlation.

Keywords Recommender system · Multi-criteria recommender system · Collaborative filtering · Similarity measure

1 Introduction

In order to select a good product from a bundle of products or make the right decision in any situation, it is usually required to process an adequate amount of information. In the era of modern technologies, it is easy to get information. For example, if a person wants to watch a movie online, many choices are available. However, when there is much information, it is challenging to decide. So here comes the problem of information overload. The recommender system helps us deal with this problem and suggest personalized information relevant to us. Many online shopping sites and other applications use recommender systems to suggest user-relevant products according to their taste. For example, Netflix recommends movies, YouTube suggests videos, and Amazon recommends clothes, books, and many other products. Most websites explicitly ask for their users' feedback about purchased items or products. Netflix asks a user usually rate a movie in the form of stars from 1 to 5. Websites also collect users' feedback by implicit feedback, which includes browsing history, clicks on pages, etc. The task of the recommender system is to predict user preference for yet unseen items based on user feedback and activities and suggest the new items relevant to the user's taste.

The film suggestion framework is the most broadly utilized application combined with online media stages, which intends to assist users with getting to favored films

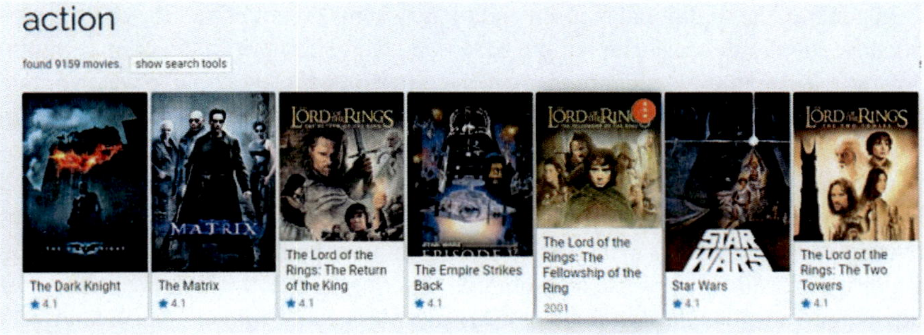

Fig. 1 An example of recommender system

keenly from a huge film library. Much work has been finished in the scholarly and modern regions in growing new film suggestion calculations and expansions. Most existing proposal frameworks depend on the collaborative filtering (CF) method. This has effectively evolved in a couple of years. It first gathers appraisals of films given by people. It then, at that point, prescribes great films to target users in view of the "similar" people with comparable preferences and inclinations before. Numerous renowned web-based sight and sound stages (e.g., youtube.com, Netflix.com, and douban.com) have been consolidated with the CF procedure to recommend media items to their customers [1].

The center of collaborative filtering is to work out similitudes among users or items. The presentation of a recommender framework relies on the precision; of how precisely the recommender framework makes predictions [2]. The forecast capacity relies upon the similarity measure used to track down comparable users. The forecast will get as worked on as much similarity measure gives improved results. Comparability can be found between users well as items [3]. Numerous scientists have proposed numerous techniques to figure similitude. Pearson Correlation and Cosine based Similarity measures are generally usually utilized.

The vast majority of the current recommender frameworks propose items to users in view of a solitary standard, i.e., the general rating. Nonetheless, users might see various models while rating a thing, so the general rating does not uncover the real user conduct. In this way, coordinating appraisals on different models into RS can catch users' way of behaving proficiently and produce more commonsense proposals for users. Figure 1 demonstrates the way that in MovieLens suggestion, users can channel films through unambiguous standards (activity).

Traditional Recommender systems compute the similarity among users based on single criteria to get a prediction for active users. They do not take into account multiple aspects of products or items. An item or product can be rated on multiple criteria such as in Restaurant system have different rating criteria in terms of its food, decor, service, and cost. The most used example is Zagat's Guide (http://www.zagat.com) which recommends restaurants in the main cities of the world using three criteria (food, services, and décor) [4, 5].

In general, the traditional recommendation systems do not use all the criteria provided by the movie's website but are based on a single criterion: the overall rating. However, a user may consider more than one criterion while deciding to watch or rate a movie. In order to give more accurate recommendations, a recommendation system can take advantage of considering multiple criteria. In multi-criteria Recommendation, different types of similarity measures (e.g., Pearson Correlation) are used, and evaluation of those measures is done on some user pairs. In this research, we will investigate how different measures suit different user pairs. So after implementing similarity measures, users use collaborative filtering to check which measure accuracy is good at: (1) Cold start users or users rated less than 5 movies. (2) When Users rated more than 5 and less than 15 movies. (3) When users rate more than 15 movies. The Objectives are:

- To propose a new model suitable and flexible for predicting ratings accurately for cold-start users, middle users (overall number of movies rated >5[R]<15), and heavy users (which rated more than 15 movies overall);
- Similarity calculations: Using each criterion in our dataset, such as visual, acting, direction, story, and overall. Similarity among users will be calculated;
- Neighbor selection: based on similarities output. Predicting similar users for active Users;
- New rating prediction using neighbors of active users;
- By train and test split method find the accuracy of similarity measure on each set of users;
- To discuss multiple variations of each proposed approach and perform an empirical analysis of approaches using a real-world dataset (Yahoo Movies);
- Evaluation of Experiments and Results.

Our motivation is that different kind of similarity measure suits different user pairs. Using threshold values, we can divide our users into three different groups. The normal distribution can specify threshold values.

Section 2 describes associated literature. Section 3 presents the problem statement and the solution outline. Section 4 descries the proposed model. Section 5 display the implementation and evaluation. Finally, Sect. 6 describes the conclusion.

2 Related Work

Recommender frameworks anticipate users' inclinations over numerous things by pooling comparable data from different users or potentially things within sight of inadequate observations [6]. Recommender frameworks are typically characterized into the accompanying classes: Collaborative sifting frameworks that endeavor to distinguish gatherings with comparable preferences to those of the user and suggest things that they have enjoyed, and Content-based recommender frameworks which utilize content data to prescribe things like those recently liked by the user [7].

Collaborative filtering utilizes various procedures, for example, memory-based and model-based approaches. In memory-based approaches, similarity among users can be found utilizing different comparability measures like cosine or Pearson. Model-based Using AI to track users' appraisals of unrated things, e.g., PCA, SVD, Matrix factorization.

Neighborhood models: Neighborhood model is a methodology utilized in CF is a calculation that uses the utilization examples of other comparable users in the system [8]. As collaborative filtering draws near, strategies depend on closest neighbors. This strategy gathers inclinations from different users and joins them to give a sensibly customized proposal to the dynamic user [9]. Collaborative filtering strategies can be gathered into the two general classes of the area and model-based methods [10]. In area-based (memory-based or heuristic-based), [11] collaborative filtering, the user thing appraisals put away in the framework are straightforwardly used to anticipate evaluations for new things. This should be possible in two ways: client-based or thing-based suggestion. Rather than neighborhood-based frameworks, which utilize the put-away evaluations straightforwardly in the expectation, model-based approaches utilize these evaluations to get familiar with a prescient model [12]. In the area-based techniques, a subset of reasonable users is chosen in light of similitudes with dynamic users, and a forecast is figured given the weighted total of their ratings [9]. The technique applies comparability measurements, for example, Pearson relationships to weight things or likeness. It utilizes all connected neighbors to register the last expectation by performing weighted normal deviation from the neighbor mean [13].

The local strategies are fundamental methods utilized right from the outset of building suggestion motors. These are the most established, at this point, most broadly utilized approaches, even today. These broadly involved approaches are famous due to their precision in producing proposals. We know that every recommender framework chips away at the idea of similarity between things or users.

Similarity Computation: Similarity in a recommender framework is tied in with tracking down things (or users, or user and thing) that are comparative. Step-by-step instructions to gauge it rely upon the recommender framework, whether content-based or collaborative filtering. On account of collaborative filtering, two things are comparable on the off chance that a specific number of individuals like or disdain them the same way.

One essential move toward the collaborative filtering calculation is to process the likeness between Users/things and select the most comparative Users [14]. The essential thought in likeness calculation between two Users u and u' is first to confine the users who have evaluated both of these things and afterward apply a similitude calculation method to decide the similarity. There are multiple ways of processing the similitude between Users [15]. Here we present two such strategies to figure likeness over multi-criteria evaluations. These are distance-based comparability and Pearson Correlation. In distance-based likeness, first, track down the distance among users and take the opposite of the distance. If the distance is more modest, users are more like one another, and in Pearson connection, there is a straight connection between users.

Zhao [16] combines four elements, client, individual interest (connected with the user and the thing's subjects), relational interest comparability (connected with user interest), relational rating conduct similitude (connected with users' evaluating conduct propensities), and relational rating conduct dispersion (connected with users' conduct disseminations)- into a bound together grid factorized system.

Multi-Criteria Recommendation Systems: A multi-criteria recommender framework can be characterized as a recommender framework that integrates inclinations into numerous aspects giving more data about the users and the things which could be successfully utilized in the suggestion process [17]. By far, most recommender frameworks, freely of their sort, utilize a solitary rule to assess everything and infer their suggestions. Evaluations on various measures then pass on more extravagant data about user inclinations. Reasonably, in frameworks where suggestions depend on the assessment of others, taking into account various models that can influence the users' perspectives might prompt more detailed proposals because a multi-criteria recommender framework permits addressing users' inclinations on a few viewpoints things.

This technique has helped recommender frameworks lately to conquer difficulties as it offers prescient precision and the extra benefit of adaptability for demonstrating different genuine circumstances. The following are the absolute most recent examination work in multi-criteria proposal frameworks connected with the issue we are additionally attempting to address:

Quite possibly, the earliest headway in recommender frameworks in light of multi-criteria appraisals was created by Adomavicius and Kwon (2007) [5]. The proposed new methods for consolidating multi-criteria evaluations in recommender frameworks. The two new methodologies presented in their work are the similitude-based approach and the collection capability-based approach. They utilized the Chebyshev distance metric to work out the comparability between the users or things and took the base of the cosine likeness metric. Since the general rating relies upon the single evaluations, they contended that a general rating is some capability of a solitary rating.

Muhammad Wasid [4] Proposed a bunching way to deal with really consolidating multi-criteria evaluations into customary recommender frameworks. Besides, they registered the intra-cluster user similitudes utilizing a Mahalanobis distance technique to make more detailed proposals. They contrasted the proposed approach and the conventional collaborative filtering-based strategy utilizing the Yahoo! films dataset.

Kermany and Alizadeh [18] presented a versatile neuro-fluffy deduction framework to find the connection between every measure and the general rating. A combination of fluffy cosine and Jaccard likenesses is additionally adjusted to work out the all-out similitude between users/films concerning the impact of co-evaluated thing set cardinality on the dependability of similarity measures.

Dietmar Jannach [19] proposed a recommender framework in view of multi-criteria evaluations. The proposed new strategies to integrate multi-criteria evaluations into recommender frameworks to work on prescient exactness. The idea of total capability proposed by Adomavicius and Kwon is also being utilized in this work. Be that as it may, they have utilized Support Vector Regression to decide the connection between the evaluations and generally appraisals.

3 Problem Statement and Solution Overview

3.1 Problem Description

The first phase in RSs is to retrieve users' or items' information as much as possible because when the system knows more about users' tastes, it will be easy to recommend movies/products according to users' tastes. Two main approaches are used in RSs content and collaborative filtering, where content-based only collects the information from the user profile and its implicit and explicit feedback. At the same time, CF determines how to recommend an item by collecting user feedback in the form of ratings for movies/products in a given domain and exploiting similarities in rating behavior amongst several users.

CF is a highly used filtering technique in a Single Criteria or multi-criteria rating system. The single criteria rating system deals only with User $\times$ Items two-dimensional space. If we consider the example of a simple movie recommender system, User Alex can rate the movie 'joker' 4 (out of 5 stars). As we know, Alex likes the movie, but he rated the movie on the bases of overall ratings. As we know, every movie has many aspects such as action, visuals, and story or movie genre can be adventure or romantic. However, in this example, we do not know why he likes the movie in the single criteria rating system. Let us consider another scenario. When we want to buy a car, we consider lots of attributes of a car, such as its color, price, and brand.

Although single criterion recommender systems are widely used, they can be extended to the multi-criteria rating system to address the mentioned issues. For the last ten to fifteen years, multi-criteria getting more attention as they capture users' tastes more accurately in multiple aspects.

Most of the previously used approaches, such as Adomavicius and Kwon [20] used to compute the similarity among users using different similarity measures, but they consider only high-ranked users. Liu [21] introduced a multi-criteria recommendation approach that is based on the clustering of users. Where he grouped the users based on criteria, he suggested that one of the criteria can be "dominant," and users are grouped according to their criteria preferences.

In summary, many approaches are discussed in Sect. 2 related to the issue of incorporating multi-criteria ratings in the recommender system. However, most of these approaches are implemented for all users similarly. However, sometimes similarity measure can be accurate for the high-ranked user, but it this possible that the similarity measure does not perform well for new or cold-start users. As discussed in Sect. 2, RSs behave differently at cold start and heavy users.

This ensures that we can divide our users into cold-start, middle, and heavy users to perform our proposed approaches on a multi-criteria rating system. We can find a solution to the problem of how to divide users into different groups. This section presents the two main approaches, Pearson correlation, and multi-dimensional distance metric, to compute the similarity among users on multi-criteria ratings.

3.2 Proposed Solution

As discussed in the problem description, when multi-criteria recommender systems are implemented using different similarity measures, they consider all users in one scenario. However, different users have a different overall movie rating. In the recommender system, some users are cold-start users who may be new to the system or have fewer overall movie ratings, and some high-ranked users rated enough movies. The recommender system is capable of providing them with good recommendations. So here in our proposed approaches, we will divide users into three groups based on their overall number of movie ratings. One group we call cold-start users has a total number of movie ratings less than 5, and the second group, called middle-level users, has a total number of movie ratings more significant than 5 and less than 15. finally, we have a heavy users group where each user rated more than 15 movies. We will check the probability of users by normal distribution so we can get an idea of which user belongs to which group of users. First, we will calculate similarity among users on each criterion, and all individual similarities will be aggregated for average or overall similarity. We will predict the new ratings for active users based on overall ratings. After applying our proposed similarity measure, we will check which similarity measure performs well in which group of users? We will evaluate the recommender system using precision, recall, and f1 score.

3.3 Single-Rated Versus Multi-Criteria Recommendation

The vast majority of recommender systems, independently of their type, use a single criterion to evaluate each item and derive their recommendations. Ratings on multiple criteria, on the other hand, convey richer information about user preferences, notably in systems where recommendations are based on the opinion of others.

Figure 2 shows this rating forecast process with a straightforward model. Accept we have 5 users u1... u5 and 5 films i1... i5. Besides, let us accept that the recommender framework requirements to appraise how much the objective user u1 would like film i5 and, as demonstrated in Fig. 2, that any remaining appraisals of various users to various films are known. Then, at that point, the conventional collaborative filtering approach finds the users nearest to u1 and who have seen film i5. For this situation, u2 and u3 appear to be fantastic counterparts for the objective user (u1) since they generally appraised the classic films similarly. As may be obvious, both u2 and u3 evaluated film i5 as 9, and the worth of target rating R (u1, i5) will be anticipated as 9. Figure 2 shows a conventional single proposal framework utilizing collaborative filtering one is our dynamic or target user u1, and all others are its neighbor's matches. Taking a gander at the rating of dynamic users and their neighbors can anticipate the rating for another thing that u1 has not evaluated. Presently let us expect what is going on in a multi-criteria rating framework. Once more, we have 5 similar users, u1... u5, and 5 films i1... i5. Once more, Now we can find in Fig. 3 rating R (u1, i5) that the rating of dynamic user u1 is obscure and should be anticipated, and,

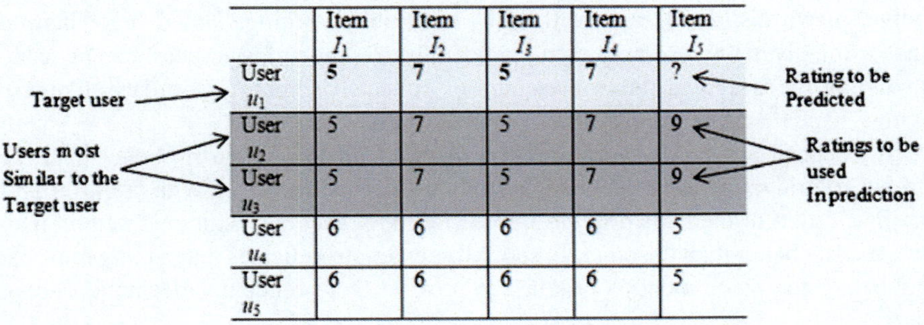

Fig. 2 Collaborative filtering in a single-criteria setting

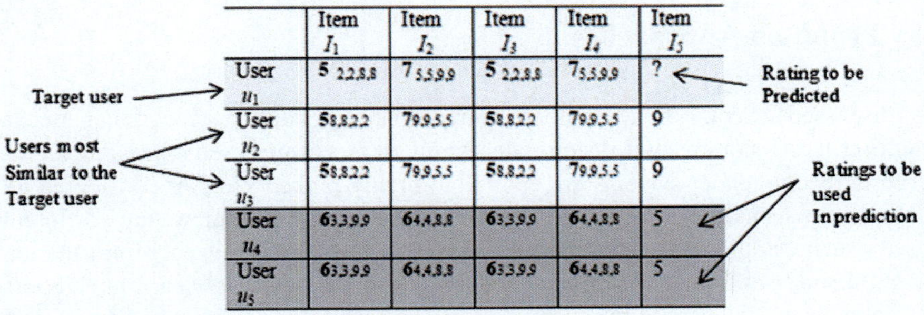

Fig. 3 Collaborative filtering in a multi-criteria setting

as displayed in Fig. 3, any remaining users' general evaluations of various users to various films are referred to and are precisely equivalent to prior. Furthermore, let us expect that every user is likewise gotten some information about the film on four explicit rules: story, acting, heading, and visuals, and that the general rating, for this situation, is a fundamental norm of the four individual models evaluations. Nonetheless, as a result of all the extra data accessible as multi-criteria evaluations, one can see that users u2 and u3 are disparate in their preferences and inclinations from user u1, even though their general evaluations for every film match impeccably. Specifically, the film perspectives that u1 despised (story and acting) were enjoyed by u2 and u3, as well as the other way around. Be that as it may, in recommender frameworks, in light of single-standards evaluations, this data would be "covered up" inside the total generally appraising, which might prompt erroneous experiences about the genuine similitude between user inclinations. user u4 and u5 appear to be significantly improved matched for user u1 in this model since their general appraisals are comparable. However, their inclinations for various film angles were practically the same also. We can see in multi-criteria that rating users u4 and u5 are like our objective users.

In light of this thought, we propose to broaden the standard collaborative filtering calculation to incorporate multi-criteria evaluations. In particular, we propose multiple ways of remembering multi-criteria rating data for computing the likeness between two distinct users sim (u, u') or two unique things sim (i, i'). Then, at that point, given the recently determined similitude, the rating forecast can be made involving the weighted total or average of collected similarity, similarly likewise with a standard collaborative filtering calculation.

It has been proved that multi-criteria recommendations perform better than the single criteria rating Recommender system in CF. However, even with multi-criteria ratings, different methods provide various levels of accuracy, and one method may fit a dataset better than the other. In short, this example indicates that adding multiple criteria to the recommender system can provide more efficient information about user taste.

4 Proposed Approaches

As discussed in Sect. 1, we will extend the traditional single-criteria recommender to a multi-criteria recommender system. Most similarity measures used in existing multi-criteria recommender systems are applied to highly ranked users or considered all users in one scenario. However, in every Recommendation system, we have different users, such as cold start users. When the system does not have much information about the user or items, it is difficult for the system to predict its nearest neighbors or items which relevant to him. However, in the case of warm users or heavy users system have enough information about users, so using that information, it is easy for the recommendation system to predict the nearest neighbors or items for the user which are relevant to him. In our dataset, we have different kinds of users where some users rated few movies, and some rated more than 100 movies. So by putting a threshold value, we can divide users into three groups: cold start users, middle-level users, and heavy users. In the implementation phase, we will divide users into three groups and apply similarity measures such as multi-dimensional distance metric and Pearson correlation.

4.1 Normal Distribution

The normal distribution is a probability distribution that is symmetric to the mean, showing that data close to the mean is frequent in occurrence compared to data far away from the mean. The standard normal distribution has two parameters mean and standard deviation. Figure 4 shows the normal distribution curve where the mean is denoted with μ and Standard deviation is denoted with σ. It is often called a "Bell Curve" because it looks like a bell. As we know, we have three groups of users such as cold start users, middle-level users, and heavy users. So we can find the probability of any user x by using a normal distribution that weather user belong to cold-start users or middle-level users or heavy users.

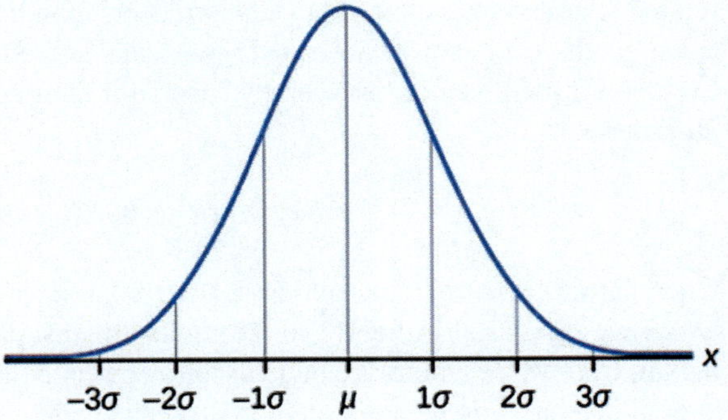

Fig. 4 Normal Distribution curve

Our cold-start users' overall ratings are 5>[R], middle users overall ratings are 5<[R] >15 and heavy users overall ratings are [R]<15.

So if we have any user x with the total number of overall movie ratings, 5 using the normal distribution, we can find its probability from which group of users it belongs. The normal distribution is given as

$$X \sim N(\mu, \sigma) \tag{1}$$

So can simply find the probability of x by
P(x/cold start users) = P (xIN(3,2)
P(x/middle level users) = P (x IN(10,5)
P(x/Heavy users)= 1−P (xIN(3,2) − P (x IN(10,2)

4.2 Multi-dimensional Distance Metrics

The concept of similarity measure is to check how much two users are similarly based on their standard ratings. In the case of a distance metric, if the distance between users becomes less, they will be more similar, and in the case of considerable distance, there will be less similarity among them.

Based on the idea, check which approach performs well on which group of users. We will calculate user similarities using our mentioned algorithm multi-dimensional distance metrics on each criterion. Once we have similarities in each criterion, we can aggregate the similarities to overall similarity and predict new ratings for the active user. The difference in our approaches is that we will divide users into different groups based on their total number of movie ratings.

We can calculate similarity using multidimensional distance metrics as we know in multi-criteria rating matric, each rating $R\ (u, i) = (r0, r1\ldots rk)$ represents a point in the k+1-dimensional space. Where $R\ (u, i)$ represents the rating of user u on item i. r, 0 represents the overall rating for the item I, $r1$ represents the first criteria ratings for the item I and so on up to k criteria. Similarity and distance are inversely related smaller the distance between two users than higher the similarity. So we can calculate the similarity between user two users in the following steps: First, we have to calculate the distance between two users for their common rated items/movies on all criteria.

$$drating(R(u, i), R(u', i)) \tag{2}$$

where R(u, i) = (r0, r1…rk) and R(u, i) = (r'0, r'1…r'k).

As we discussed in Sect. 2, three multi-dimensional distance (Euclidean, Manhattan, Chebyshev) metrics can be used to compute the distance between users. The overall distance can be computed or aggregated as

$$d_{user}(u, u') = \left(\frac{1}{|I\left(u, u'\right)|}\right)\left(\sum_{i \in I(u,u')} drating\ (R(u, i),\ R(u', i))\right) \tag{3}$$

I(*u*, *u'*) denotes the set of items that both *u* and *u'* have rated. Alternatively, this is simply the overall or average distance between two users. By taking the inverse of distance, we can get the similarity between two users.

$$Sim\ (u,u') = \left(\frac{1}{1 + dUser(u, u')}\right) \tag{4}$$

Note that this definition of similarity has desired range properties, i.e., the similarity will approach 0 as the distance between two users becomes larger, and it will be one if the distance is zero (users are identical). 1 added to the distance to avoid the divided by zero error.

4.3 Pearson Correlation

Pearson similarity measure is based on how much the ratings by common users for a pair of items deviate from average ratings for those items.

As mentioned, some details about Pearson correlation in Sect. 2 is widely used collaborative filtering for measuring the similarity between user to user or item to item. Pearson correlation can be applied to multi-criteria rating scenarios. Using Pearson correlation, first, we calculate the similarities between users on each criterion. Let's consider that each rating user u will give to an item i consist of "Overall' rating r_0 and k multi criteria ratings $r_{1,}\ r_{2...}r_k$.

$$R(u, i) = (r0, r1, \ldots, rk) \tag{5}$$

Now, $k + 1$ similarity estimates can be obtained by using a similarity measure to compute the similarity between user *u* and *u'*: Sim_0(*u*, *u'*) is similarity between two

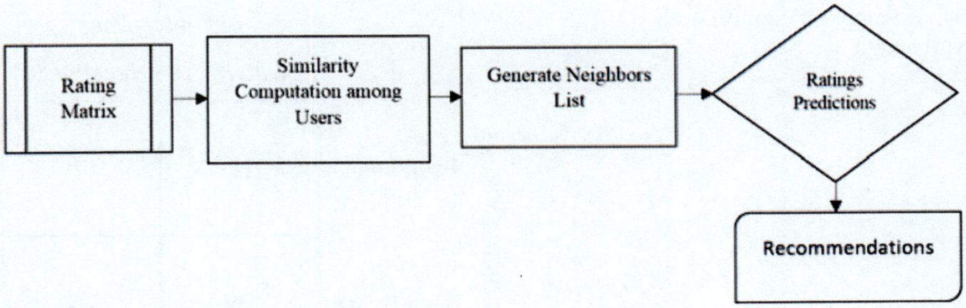

Fig. 5 Process of rating prediction

user on overall ratings, *sim*1(*u*, *u'*) represents the similarity between two users on first criteria, *sim*2(*u*, *u'*) is the similarity between two users on the second criteria, and so on for the rest of the criteria. We can aggregate the individual similarities by using Eq. 2.

$$\left(\text{sim}_{\text{avg}}\left(u, u'\right) = \frac{1}{k+1}\sum_{i=0}^{k}\text{sim}_i\left(u, u'\right)\right) \tag{6}$$

4.4 Ratings Prediction

Rating prediction is a recommendation system task aiming to predict the user's ratings for those items/movies he has not rated yet. Once the overall similarity is calculated, it is easy to predict similar users for the active user. In the implementation phase, first, we will calculate user-to-user similarity based on their standard rated movies on each criterion, then aggregate the overall similarity. As we will get similarity on each criterion, we can get the aggregated similarity score by taking an average of all individual similarities. Based on aggregated similarity score, we will predict the neighbor's list of the active user.

After generating the neighbor's list for the active user, we predict the rating for new movies on the bases of previous user ratings and similar users. Rating perdition will be done on the bases of users' overall ratings. New rating prediction will be made for the movies active users have not rated yet.

Figure 5 shows the process of rating prediction for the user. First, we have a rating matrix where users rated movies on four criteria with overall ratings. The next step is to calculate similarities among users User × User similarity. When we get each criteria similarity, we can average or aggregate similarity. Similar users can be extracted using this aggregated similarity. New movies which users have not rated can be presented to the user by predicting their ratings.

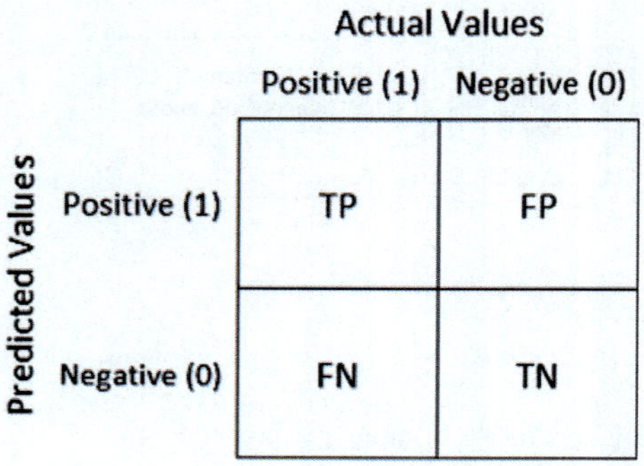

Fig. 6 Sample of confusion matrix table

4.5 Model Evaluation

Model Evaluation is a fundamental piece of the model advancement process. The objective of multi-criteria recommender frameworks is to find things that amplify every client's utility, similar to the single-rating recommender frameworks. Subsequently, one of the suggestion frameworks' fundamental objectives is to anticipate everything's general rating for every user because the framework, at last, requires to look at the things in light of their general evaluations and prescribe the best things to the users. The distinction between single-rating and multi-criteria rating frameworks is that the last option has more data about the users and things, which can be utilized in the proposal cycle.

Several metrics, such as precision and recall, help us interpret the model's accuracy and choose the best model. We will use the concept of a confusion matrix. A confusion matrix is a table often used to describe the performance of a model on a set of test data for which the actual values are known.

Figure 6 shows the confusion matrix table where we have predicted and actual values. If predicted values and actual values both have a value of 1 then results will be True positive, and if predicted ratings do not satisfy the threshold value, they will be treated as negative. The concept of confusion matrix classifies things as belonging to either class 1 (we will call these positives) or class 0 (call them negatives). Each data point, therefore, belongs to one of four classes:

(1) True positives: predicted class 1, really belongs to class 1
(2) True negatives: predicted class 0, really belongs to class 0
(3) False positives: predicted class 1, really belongs to class 0
(4) False negatives: predicted class 0, really belongs to class 1

In this section, in the beginning, we learned why a multi-criteria recommender system is better than a traditional single-criteria recommender by considering an example. Normal distribution approach is discussed for finding the probability of user checking any specific user belongs to which group.

Further, we discussed multi-dimensional metrics, Pearson correlation similarity measure, and how individual similarities will be aggregated. At the end of this section, we learned about rating prediction and evaluation matrix.

5 Implementation and Evaluation

This section focuses on the implementation details of the recommender system. We did the dataset pre-processing and then implemented the proposed model similarity measure. We discussed rating prediction, and finally, model evaluation is explained.

5.1 Data Analysis and Pre-processing

5.1.1 Dataset Description

The dataset [22] is available for free download from the yahoo movies website webscope.sandbox.yahoo.com which contains 6078 total users and 976 movies. This is a multi-criteria dataset where each user rated a movie on four criteria, visual, story, acting, and direction, with overall ratings. The total number of ratings available is 62156, where average user ratings equal 9.6193, and average movie ratings are 9.0112. Dataset is in CSV file which directly loaded to Jupyter notebook by load function. The dataset [22] is available for free download from "webscope.sandbox.yahoo.com" containing 6078 total users and 976 movies.

Table 1 contains a sample of the yahoo movies dataset where we can see two users rated different movies on four multiple criteria and overall ratings. The train and test split ratio are 70/30, where 70% for training the model and 30% test the model. The training set must be separate from the test set.

Table 1 Sample of the numerical representation of the original dataset

User ID	Movie ID	Acting k_1	Story k_2	Direction k_3	Visual k_4	Overall k_0
101	1	13	5	6	5	5
	3	9	10	10	10	12
	5	8	10	11	10	11
102	3	13	12	12	10	13
	5	6	7	11	12	12
	6	6	9	7	5	8

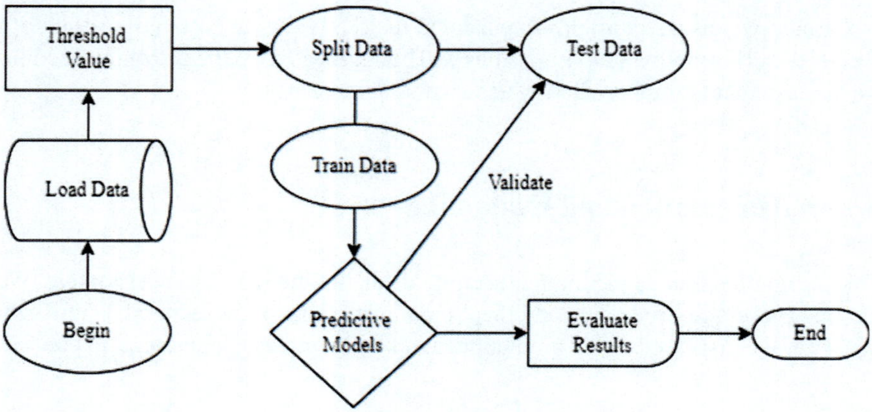

Fig. 7 The model architecture

5.2 *Architecture of the Proposed System*

We will divide our users into three different groups, such as cold start users, middle-level users, and heavy users, based on their overall ratings using specific threshold values. The procedure of similarity computation and rating prediction will remain the same for all. We will first predict the user's ratings on the training dataset for the movies he has not rated yet. However, for evaluating our approaches, we will use a test dataset to check the accuracy of our approaches on each group of users. Figure 7 shows the system architecture and how the system is implemented and achieves the functionalities proposed. The similarity computation is done using multi-dimensional distance metrics and Pearson correlation. The similarity analysis helps in identifying the neighbors and generating the prediction. The effectiveness of the methods is observed by precision, recall, and f1 score from the test cases data, which is actual data split into 70% for learning and 30% for testing. As our dataset ratings scale is 1 to 13 so, for test evaluation in the confusion metric, we selected a threshold value is 9.

5.2.1 Rating Distribution

Different datasets have rating scales, such as 1 to 5 stars or like s and dislikes. In yahoo movies, the multi-criteria rating scale is 1 to 13, where users can rate the movie from 1 to 13 scale.

A rating scale for each criterion, Visual, acting, story, and direction, is the same 1 to 13 scale. Figure 8 shows rating distribution details of the yahoo movies dataset, which we can see from counting the total number of ratings on each scale.

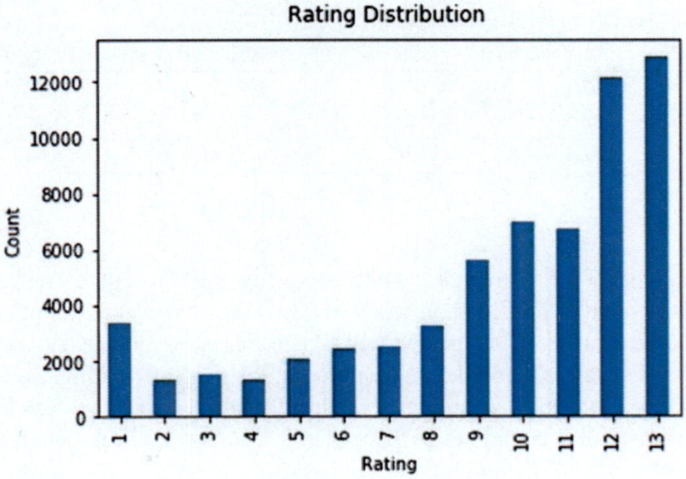

Fig. 8 Rating distribution of Yahoo movies dataset

5.3 Evaluation Metrics

Did we divide our users into three groups based on their overall ratings to check which algorithm performs well in which group of users? By setting the threshold value in the cold-start users' group, we end up at 1000 users and 710 movies which every user rated less than 5 movies. In the second group, we have middle-level users where every user rated more significant than 5 and less than 15 movies. We end up with a total of 550 users and 500 movies. Moreover, in the final group, we got heavy users where every user rated more than 15 movies overall. In this group, we ended up with a total of 100 users and 700 total movies.

The recommender system can be evaluated by using the confusion metric. We used the test dataset to evaluate the results by calculating the precision, recall, and F1 score.

5.3.1 Model Evaluation

Now we used the test data to evaluate multi-dimensional distance metrics and the Pearson method used for each group of users. Several metrics, such as Precision and Recall, help us interpret the model's accuracy and choose the best model. We will use the concept of a confusion matrix. A confusion matrix is a table often used to describe the performance of a model on a set of test data for which the true values are known. In model evaluation, we use test data which is 30% of actual data. We converted the actual and predicted rating lists into 0 and 1.

The source code of the evaluation model method is below where we specified the threshold value 9 (i.e., 1 to 13 scale rating) for the confusion metric where predicted

Table 2 Cold start users evaluation results

Measures	Manhattan (%)	Euclidean (%)	Chebyshev (%)	Pearson (%)
Precision	67.1	67.4	67.3	60.3
Recall	78.5	78.5	78.5	73.7
F1 Score	72.4	72.5	72.4	66.43

rating higher than 9 will be treated as 1 or true positive and below than 9 will be treated as 0 or false positive. This method works the same for the evaluation of cold-start, middle, and heavy users. This method will take test data, predict the list and evaluate the model as lists are already defined for storing the predicted rating. So using this method, we will calculate the Precision, Recall and F1 score. We calculated the Precision and recall accuracy for each similarity measure to check which similarity measure performs well in which group of users.

In model evaluation, we only consider the test dataset, which is 30% of the actual dataset. We evaluated the results of each similarity measure on each group of users.

5.3.2 Cold Start Users

In the yahoo movie, the multi-criteria rating dataset's total number of users is 6078, and the total number of movies is 976. However, in cold-start users, we consider users who rated less than 5 movies overall. By specifying [R] < 5, we ended up with 1000 Users and 710 movies. After dividing the dataset into train and test, Multi-dimensional distance metrics and Pearson correlation are implemented on the multi-criteria dataset. After finding the similarity among users, we predicted the new ratings for users u for the movies they have not rated yet. Once we saw that our model was predicting the new ratings, we provided a test dataset to our model and calculated the Precision, Recall, and F1 Scores for all. Results are shown in the table below, where Euclidean distance is performing well compared to other similarity measures such as Manhattan, Chebyshev, and Pearson. Table 2 presents cold start users evaluation results.

5.3.3 Middle Level Users

For middle-level users, our threshold value is 5> [R]< 15, which means that in this group, we will consider the users who rated overall greater than 5 movies and less than 15. In this group, we have 550 Users and 722 movies-the average number of ratings per user is 8, and the average number of ratings per movie is 6. After dividing the dataset into train and test, Multi-dimensional distance metrics and Pearson correlation are implemented. We calculated the Precision, Recall, and F1 Score for the test dataset for all. Results are shown in the table below, where Euclidean distance is

Table 3 Middle level users evaluation results

Measures	Manhattan (%)	Euclidean (%)	Chebyshev (%)	Pearson (%)
Precision	64.6	64.2	64.7	67.4
Recall	76.5	76.7	76.5	71.4
F1 Score	69.9	69.9	70.1	70.4

Table 4 Heavy users evaluation results

Measures	Manhattan (%)	Euclidean (%)	Chebyshev (%)	Pearson (%)
Precision	64.3	64	64.1	64.1
Recall	71.5	71.5	71	63
F1 Score	67.7	67.5	67.4	63.5

performing well compared to other methods. Table 3 presents middle level users evaluation results.

5.3.4 Heavy Users

For heavy users, we took the users from our dataset, which rated more than 15 movies overall. In this group, we have 140 users and 776 movies. In this group, the average user rating is 9.69, and the average movie rating is 9. After applying the proposed approaches to multi-dimensional distance metric and Pearson correlation. Model Evaluation is done, and we get the Precision, Recall, and F1 score for each approach used. The table below shows the results where we can see that overall, Manhattan distance performs well compared to other methods. In Table 4, heavy user evaluation results are presented. We can say that when there are heavy users in the recommendation system, then Manhattan distance similarity can be used for better user × user similarity and predicting or recommending new items to the user based on its similarities users.

5.4 Evaluation Results

This section presents the evaluation results of all similarity measures used on a different group of users. On the bases of precision, the ability of a classification model to identify only the relevant data points. Precision is defined as the number of true positives divided by the number of true positives plus the number of false positives. False positives are cases the model incorrectly labels as positive that are negative.

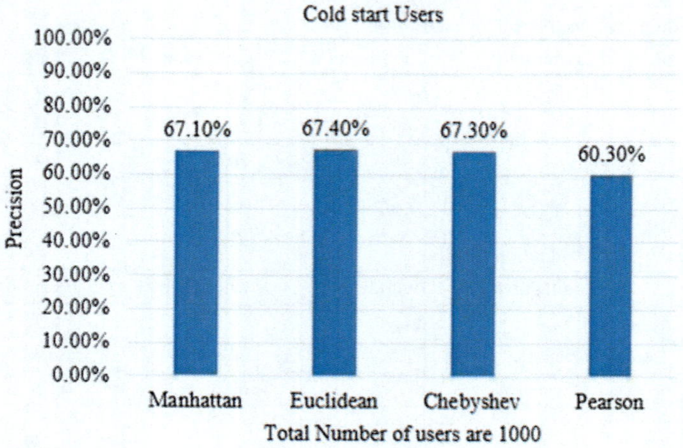

Fig. 9 Similarity measures results of cold start user

As discussed in Sect. 1, we divided users into three groups: cold start users, middle-level users, and heavy users. Four similarity measures were applied to these three groups of users. For each user group, every similarity measure works in its way. We noticed that different similarity measure suits different pairs of users. Here in the form of a graph, results are presented, which show the results in which group of the user whose similarity measure is performing well.

5.4.1 Cold Start Users

The recommendation can be made when the system has enough information about users or items. However, in the case of a cold start, users' systems have less information, so it is difficult for the recommender system to predict new ratings for them or give them a recommendation. In our case, we considered those users cold-start users who rated less than 5 movies.

Figure 9 shows the similarity measures results when we have cold-start users. We took 1000 users and 710 movies in the cold start users group. Each user rated each movie on four different criteria with the overall rating. According to precision results, Euclidean distance similarity performs better than the other similarity measures.

In short, we can say that when there are cold-start users in the recommendation system, then Euclidean distance similarity can be used for better user × user similarity and predicting or recommending new items to the user based on its similar users. Pearson correlation is not performing well in cold-start users.

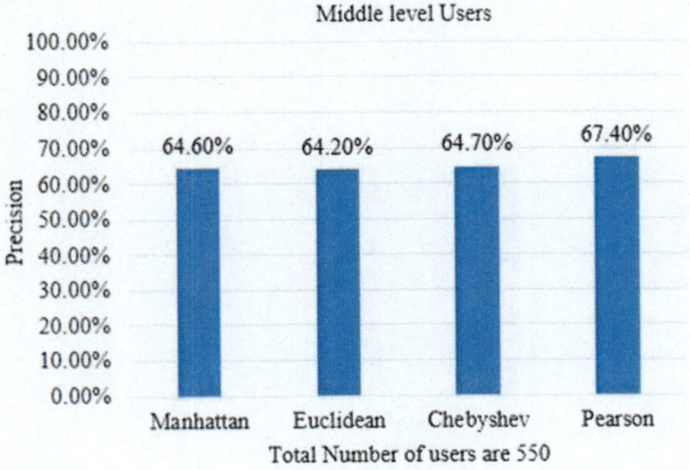

Fig. 10 Similarity measure results of middle level users

5.4.2 Middle Level Users

The middle users' overall movie rating range is that they rated movies greater than 5 and less than 15. Our total number of users in this group is 550, with the total number of movies 722.

We implemented four different similarity measures on this group of users to calculate the similarities among users. Distance metrics and Pearson correlation similarities are calculated among users. One base of similarity is that we predict new ratings for users. The evaluation of the used similarity measure is done on the test dataset to check which similarity measure is performing well at middle-level users. In short, we can say that when there are middle-level users in the recommendation system, then Pearson correlation similarity can be used for better user × user similarity and predicting or recommending new items to the user on the bases of its similar users. Euclidean distance is not performing well in middle users compared to Pearson correlation.

Figure 10 shows the results of each similarity measure used in middle-level users. The Pearson correlation results are dominant here compared to the other similarity measures. So we can say that at middle-level user pairs which rated movies 5>[R] <15, Pearson correlation similarity can be used for better recommender system performance.

5.4.3 Heavy Users

We applied a threshold value to get users from a dataset that rated more than 15 movies. We ended up with a dataset containing 140 users and 776 movies. Each movie

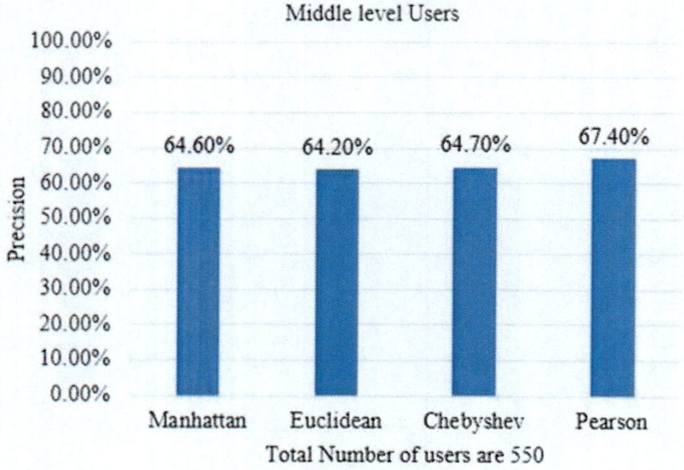

Fig. 11 Similarity measure results of heavy users

is rated by each user on four multi-criteria such as acting, visual, story, and direction, and also has overall ratings. Figure 11 shows the similarity measure results of heavy users. We can say that when there are heavy users in the recommendation system, then Manhattan distance similarity can be used for better user × user similarity and predicting or recommending new items to the user based on its similar users. Pearson correlation is not performing well in middle users compared to Manhattan distance.

6 Conclusions

In this paper, we studied extending the traditional single-criteria recommender system to a multi-criteria recommender system. In a related study, Sect. 2, we studied approaches used for similarity measuring in a multi-criteria recommender system. We analyzed that approaches used earlier only consider high-ranked users, and they do not give us an idea of how similarity measures perform when we have cold start users in a multi-criteria recommender system. We evaluate the performance of our similarity measures by using yahoo movies' multi-criteria rating. This is a dataset of user-movie ratings collected from Yahoo Movies, https://webscope.sandbox.yahoo.com/ containing 62,156 ratings in the range 1–13, where each user rated movie four multiple criteria and overall rating of 976 movies rated by 6,078 users. In this work, we divided our dataset users into three different groups. We studied that using the normal distribution probability. We have users who rated less than 5 movies in the cold start users group. We implemented four different similarity measures on this group of users dataset and did an evaluation. According to our experimental results, Euclidean distance is performing well compared to the other similarity measures used. In middle-level group users, only those users are taken into account that rated overall more than 5 and less than 15 movies. Evaluation of Similarity measures indicates that the Pearson correlation method performs well compared to multi-dimensional distance metrics such as Manhattan, Euclidean, and Chebyshev. After using four different Similarity measures such as Manhattan distance, Euclidean, Chebyshev, and Pearson correlation, we analyzed that Manhattan distance performs well compared to the rest of the similarity measures. We used Multi-dimensional distance metrics and Pearson correlation to find the similarity among users on multi-criteria ratings. Our empirical analysis of approaches using a real-world dataset shows that different similarity measures suit different user pairs. We have done recommender system evaluation using a different measures such as precision, recall, and f1 score.

Future work can be conducted on other datasets, like the restaurant recommender system, where context location users can be added to the system, and the system can be computed on multi-criteria ratings. The concept of cold-start users in a multi-criteria recommender system has not been studied as much compared to single criteria ratings, so further cold-start users' issues can be explored.

Declarations The authors state that they have no known competing financial interests or personal ties that could appear to have influenced the work disclosed in this study.

References

1. Z. Wang, X. Yu, An improved collaborative movie recommendation system using computational intelligence. J. Visual Lang. Comput. 607–675 (2014)
2. Z. Tan, L. He, An efficient similarity measure for user-based collaborative filtering recommender systems inspired by the physical resonance principle. IEEE Access 27211–27228 (2017)
3. A. Agarwal, Similarity measures used in recommender systems: a study. Int. J. Eng. Technol. Sci. Res. 619–626 (2017)
4. M. Wasid, M. Ali, An improved recommender system based on multi-criteria clustering approach clustering approach. Procedia Comput. Sci. 93–101 (2018)
5. G. Adomavicius, New recommendation techniques for multi-criteria rating systems. IEEE Intell. Syst. 48–55 (2007)
6. B. Dai, A. Qu, Smooth neighborhood recommender systems. J. Mach. Learn. Res. 1–24 (2019)
7. L.M. De Campos, J.F. Huete, M. Fernández-luna, Combining content-based and collaborative recommendations: a hybrid approach based on Bayesian networks. Int. J. Approximate Reasoning 785–799 (2010)
8. A. Gogna, A. Majumdar, A comprehensive recommender system model improving accuracy for both warm and cold start users. IEEE Access 2803–2813 (2015)
9. J. Antony, K. Akshay, Collaborative filtering using Euclidean distance in recommendation engine. Indian J. Sci. Technol. 37–42 (2016)
10. Valcarce D, Parapar J, Finding and analysing good neighbourhoods to improve collaborative filtering. Knowl. Based Syst. 193–202 (2018)
11. H. Chandrashekhar, B. Bhasker, Personalized recommender using entropy based collaborative filtering. J. Electron. Commer. Res. 214–237

12. B. Amini, R. Ibrahim, Discovering the impact of knowledge in recommender systems: a comparative study. Int. J. Comput. Sci. Eng. Surv. (IJCSES) (2011)
13. P. Andrei-Cristian, Implementation of a recommender system using collaborative filtering. Studia University Babes-Bolyai, informatica **55**, 75–84 (2010)
14. M. Schwarz, M. Lobur, Analysis of the effectiveness of similarity measures for recommender systems, in *14th International Conference The Experience of Designing and Application of CAD Systems in Microelectronics (CADSM)* (Lvov, 2017), pp. 275–277
15. B. Sarwar, K.J. George, R. John, Item-based collaborative filtering recommendation algorithms. ACM 1–9 (2001)
16. G. Zhao, X. Qian, X. Xie, User-service rating prediction by exploring social users' rating behaviors. IEEE Trans. Multimedia 496–506 (2016)
17. X. Gou, Z. Xu, Multiple criteria decision making based on distance and similarity measures under double hierarchy hesitant fuzzy linguistic environment. Comput. Indus. Eng. 516–530 (2018)
18. N. Kermany, S.H. Alizadeh, A hybrid multi-criteria recommender system using ontology and neuro-fuzzy techniques. Electron. Commer. Res. Appl. **21**, 50–64 (2017)
19. D. Jannach, F. Gedikli, Accuracy improvements for multi-criteria recommender systems, in *Proceedings of the 13th ACM Conference on Electronic Commerce* (2012), pp. 674–689
20. J. Wei, K. Chen, J. He, Collaborative filtering and deep learning based recommendation system for cold start items. Expert Syst. Appl. **69**, 29–39 (2017)
21. L. Liu, L. Xu, N. Mehandjiev, Multi-criteria service recommendation based on user criteria preferences, in *Fifth ACM Conference on Recommender Systems (RecSys 2011)* (Chicago, IL, USA, 2011), pp. 77–84
22. Yahoo movies dataset, https://webscope.sandbox.yahoo.com/

Speeding Up and Enhancing the Hyperspectral Images Classification

Dalal AL-Alimi, Mohammed A. A. Al-qaness, and Zhihua Cai

Abstract Hyperspectral images (HSI) are adjacent band images generally utilized in many different real-life applications. HSI is rich in spatial and spectral information, providing very accurate information. However, their high dimensionality is the biggest challenge in the classification process time. This study provided the solution to handle the high HSI dimensionality and speed up its classification operation with increased accuracy. The model of this study used the Exponential linear units (ELU) activation function to solve the problems of dying ReLu and the vanishing gradient. Moreover, it used QPCA to enhance dimensional reduction. The proposed model is ELUSNet. It provided the best results and time compared to five other known models with two different HSI datasets.

Keywords ELU · Hyperspectral image · Classification · CNN · Hybrid model · ReLu

1 Introduction

Recently, hyperspectral remote sensing (HRS) has been the primary way to observe the earth and discover space. HRS provides hyperspectral images (HSIs), including hundreds of spectral bands. It is a 3-D image (cube) of the same sight with different electromagnetic wavelengths, and each band in this image is linked with a particular wavelength. Each band acts as a fingerprint for the identification of its material type. The rich information in the HSIs is used in many applications and fields like

D. AL-Alimi · Z. Cai
School of Computer Science, China University of Geosciences, Wuhan 430074, China
e-mail: dalal@cug.edu.cn

Z. Cai
e-mail: zhcai@cug.edu.cn

M. A. A. Al-qaness (✉)
College of Physics and Electronic Information Engineering, Zhejiang Normal University, Jinhua 321004, China
e-mail: alqaness@zjnu.edu.cn; alqaness@yahoo.com

M. Abd Elaziz et al. (eds.), *International Conference on Artificial Intelligence Science and Applications (CAISA)*, Advances in Intelligent Systems and Computing 1441,
https://doi.org/10.1007/978-3-031-28106-8_4

mineralogy, agriculture, hydrological science, geological science, target detection, and military applications [1, 2].

On the other hand, the HSI is known for its variances; it has a large dataset, many classes with different sizes, and a complex data distribution. Moreover, the hundreds of bands of HSI have redundant data, wide dimensionality, mixed pixels, and more. All these issues complicate both reduction and classification and increase the processing time. The development of deep learning in computer vision (CV) has provided many methods to enhance the HSI classification, like the convolutional neural network (CNN) with its different dimensions (1D, 2D, and 3D). The 1DCNN obtains the spectral information from the HSI, and the 2DCNN is the best method to obtain the spatial data in each band [3–5]. Because the accuracy of HSI classification depends on spatial and spectral information simultaneously, the 1DCNN and 2DCNN are insufficient to extract well the HSI features.

In order to enhance the classification of the HSI very well, extracting spectral-spatial information is needed. The 3DCNN extracts the features from the HSI data cube and obtains higher performance than the 2DCNN. Because of the features offered by the 3D CNN, most studies converted to using it [6–8]. Due to the size of the HSI, while training, the 3DCNN exhausts the RAM and the time. Thus, many studies tried to speed up the extraction of the spectral-spatial features by combining the output of the 3D and 2D CNN layers. Yu et al. [9] used a 2D-3D CNN scheme, the 2D CNN to extract features, and the 3D CNN to utilize band co-relation data using a smaller kernel. The 2D-3D CNN model's evaluation results demonstrated its effectiveness in feature selection and increasing classification accuracy. Due to dying nodes using ReLU as an activation function (AF) in the higher layers, this study [10] used a hybrid model of 3D-2D CNN to extract the features with non-monotonic AF and compared those with the results of the ReLu. The meta-learner technique was used to classify the HSI in [11] to get the advantages of the deep and shallow networks and observe high accuracy for all the different class sizes. Different kernel sizes were used to adapt to the differences in the HSI classes [11–14].

The ReLu AF has been extensively utilized in CV HSI classification algorithms. The AF was considered a detector to generate the feature map in each layer of CNN. The ReLU function aims to solve the problem of vanishing gradient (VG) in the deep network [1, 2]. In deep learning, it helps reduce the density and speed up the model computation. On the other hand, converting the negative gradient into zero leads the corresponding nodes to stop having any influence on the learning process, called dying ReLU. Batch normalization (BN) is also an excellent way to solve the overfitting and VG problems in deep learning methods, but it slows down the training operation [11]. Exponential linear units (ELU) AF provides a good, smooth, and speedy process compared to BN and ReLu AF [15–18].

This work uses a hybrid 2-3DCNN model with ELU AF and different kernel sizes to extract more HSI features and speeds up the classification process. Moreover, it uses quantile transformation principal component analysis (QPCA) [11] to reduce the dimension of the HSI and enhance the data distribution, which will correct the spatial information and enhance the final accuracy.

2 The Framework

The whole framework of this study is in Fig. 1. The model of this study includes three layers of 3DCNN, followed by a layer of 2DCNN, and then two fully connected (FC) layers. The activation function in all these layers is ELU, as seen in Table 1.

Table 1 shows that each layer of CNN has a different kernel size because of the differences in the HSI classes. Because the 3DCNN takes a long time, dropout layers are used to avoid overfitting and reduce the training time. Since learning can be made faster by centering the activations at zero, ELU is used as an activation function to achieve mean zero rather than BN. Due to the scale of the HSI and backpropagation in deep learning models, dying ReLu probably occurred during training. Thus, that will affect the accuracy and slow down the process. So, ELU is an excellent alternative to ReLU since it decreases the bias shift by pushing the mean activation toward zero. The equation of the ELU is as follows:

$$elu(x) = \begin{cases} \exp(x) - 1 \; if \; x \leq 1 \\ x \; if \; x > 1 \end{cases} \tag{1}$$

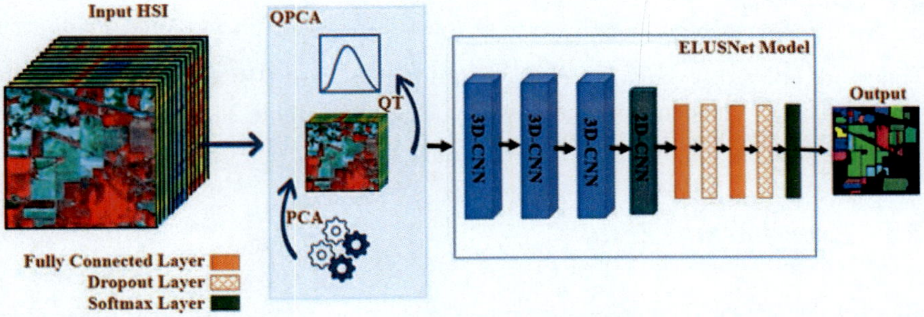

Fig. 1 The suggested framework uses the QPCA to reduce dimension and normalize distribution before passing the input data onto the ELUSNet model to categorize the extracted features

Table 1 The information of the used model and its layers

Layer	(Filters) (kernel_size) (AF)
3DCNN	(8) (3, 3, 7) (ELU)
3DCNN	(16) (3, 3, 5) (ELU)
3DCNN	(32) (3, 3, 3) (ELU)
2DCNN	(64) (3, 3) (ELU)
FC	(256) (–) (ELU)
Dropout	40%
FC	(128) (–) (ELU)
Dropout	40%
FC	(No. of classes) (–) (softmax)

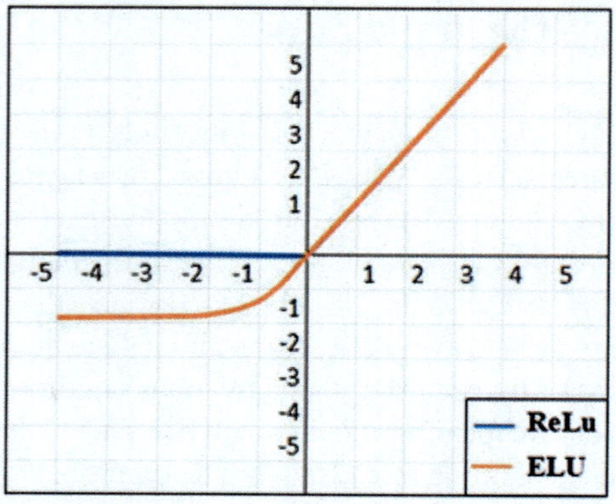

Fig. 2 The different outputs of the ReLu and ELU activation functions

where x is the input data.

Figure 2 shows the differences between ReLu and ELU AFs. The framework structure is like hybrid spectral CNN in [14] but with ELU AF, called ELUSNet.

2.1 Datasets

This work utilized two distinct datasets, the Indian Pines (IP) and Kennedy Space Center (KSC).[1] The details of these two datasets are in Table 2.

2.2 Experimental Results

Figure 1 depicts the data processing pipeline. The input data is first passed through the QPCA to minimize the data dimension, choose only the top informative bands, and enhance the data distribution. Before feeding the data into the ELUSNet classification model, the input HSI of the two used datasets (IP and KSC) is reduced by QPCA into 15 features (bands). All of these will contribute to facilitating and accelerating the classification process. Then the minimized data was fed into the ELUSNet classification model. The first three series layers of 3DCNN worked to extract the spectral-spatial features at the same time. Then these spectral-spatial features passed into 2DCNN to further improve the spatial features. The output of the 2DCNN was then transmitted into two FC layers and two dropout layers. The feature extraction was improved considerably by the FC. Processing time was slashed thanks to dropout layers and the existing 2DCNN. The VG issue was avoided, and the extraction was sped up and made smoother by using the ELU activation function in all ELUSNet model layers. Finally, the classification results were obtained by the softmax layer.

[1] https://www.ehu.eus/ccwintco/index.php/hyperspectral_remote_sensing_scenes.

Table 2 The information of the used datasets of this study, IP and KSC. **a** is the IP dataset with its ground truth, and **b** is the KSC dataset with its ground truth

Dataset	Sensor	Band numbers	Spatial dimensions	Spatial resolution	Classes number
IP	AVIRIS	200	145 × 145	20 m	16
KSC		176	512 × 614	18 m	13

(a)

(b)

The number of epochs and batch size are 100 and 360, respectively, and the learning rate is 0.01. Each dataset was split into 20% training and 80% testing. The size of the window was 15. All of the experiments were run on a 128 GB RAM machine with an 89 GB GPU and Windows 10 64-bit.

Moreover, all the experiments were repeated many times, and the accuracy of the final result was evaluated by the Kappa coefficient (KA), overall accuracy (OA), and average accuracy (AA) in each model used in this study. The output of the ELUSNet model was compared with five different other HSI classification models. They are multilayer perceptron (MLP), 1DCNN, 2DCNN, 3DCNN [13], and hybrid spectral CNN (SNet) [14].

The input data of 1DCNN and MLP was reduced into 30 features and the others into 15 bands. The MLP and 1DCNN work to extract the spectral information, and the 2DCNN extracts the spatial information from the HSI. The other models (3DCNN, Sent, and ELUSNet) worked to improve the extraction of spectral-spatial information. Tables 3 and 4 show that the ELUSNet model achieved the best accuracy because QPCA enhanced the data distribution and ELU normalized the training without losing the active nodes of the CNN layers.

Because the data distribution of the KCS dataset is quite complicated and the other models employed the PCA to minimize the dimension, these models achieved inferior accuracy compared to ELUSNet with QPCA. The QPCA shows very remarkable results compared to the others, Table 4. Moreover, the KSC dataset's data distribution is more complex than the IP dataset [11], so the IP dataset yielded more accurate results than the KSC dataset, Tables 3 and 4.

In addition, the QPCA normalized the input data, and the ELU AF was utilized, speeding up the ELUSNet processing, as seen in Fig. 3. The ELUSNet model got

Table 3 The results of all used models for the IP dataset

	MLP	1DCNN	2DCNN	3DCNN	SNet	ELUSNet
KA (%)	46.23 ± 4.27	64.38 ± 2.83	75.68 ± 2.2	94.52 ± 1.34	95.1 ± 7.09	**98.28 ± 0.25**
OA (%)	53.47 ± 3.56	69.15 ± 2.39	78.69 ± 1.94	95.19 ± 1.18	95.69 ± 6.26	**98.49 ± 0.22**
AA (%)	42.24 ± 5.22	55.78 ± 3.97	68.58 ± 3.81	90.54 ± 3.43	94.61 ± 6.29	**97.63 ± 2.19**

Table 4 The results of all used models for the KSC dataset

	MLP	1DCNN	2DCNN	3DCNN	SNet	ELUSNet
KA (%)	32.95 ± 13.46	44.42 ± 1.7	48.84 ± 9.36	81.79 ± 6.64	84.63 ± 2.87	**96.59 ± 1.23**
OA (%)	40.45 ± 12.26	51.8 ± 1.47	54.61 ± 8.01	83.56 ± 6.11	86.24 ± 2.55	**96.94 ± 1.11**
AA (%)	27.3 ± 8.78	32.67 ± 1.37	43.04 ± 8.66	81.28 ± 4.87	80.64 ± 3.25	**95.11 ± 1.64**

the lowest training time for the two datasets. In the testing time, the ELUSNet model was faster than SNet, which uses ReLu and PCA. The output of each model for the IP and KSC datasets is presented in Figs. 4 and 5. Also, Fig. 6 shows the accuracy during the training time for all models and the two datasets. The ELUSNet model provided the stablest and smoothest operation.

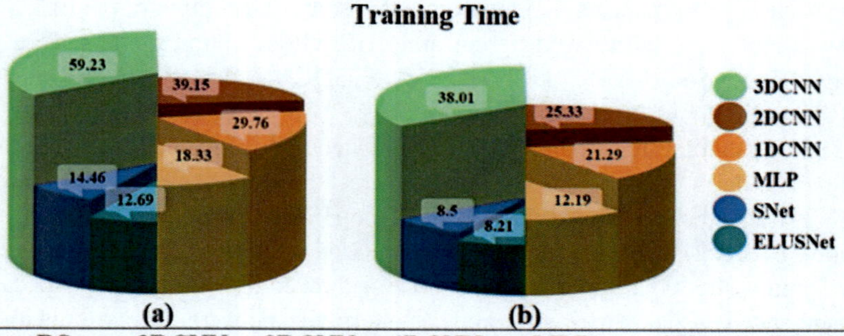

Test Time DS	3DCNN	2DCNN	1DCNN	MLP	SNet	ELUSNet
IP	1.12	0.87	0.75	0.43	1.18	0.93
KSC	0.73	0.61	0.52	0.28	0.55	0.52

Fig. 3 The performance time in all used models, **a** is the training time o the IP dataset and **b** is for the KSC dataset

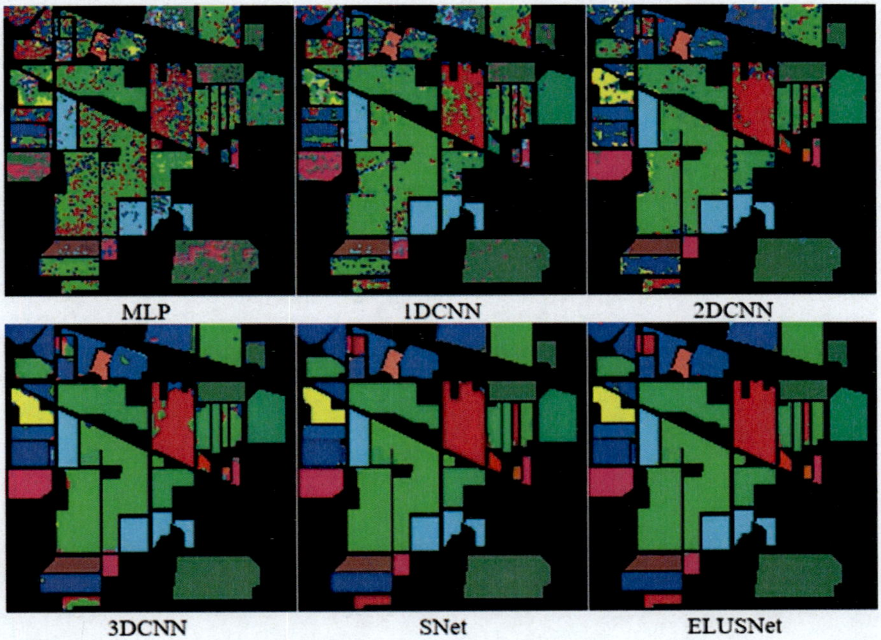

Fig. 4 The prediction of the six models for the IP dataset

3 Conclusions

The efficiency of using the ELU activation function and QPCA in the HSI classification models was demonstrated in this work. The QPCA was employed in this study to normalize the reduced dimensionality of the input HSI before it was fed into the classification model, which solved the majority of HSI complexities. ELU activation functions were implemented in all ELUSNet model layers in order to avoid VG issues and enhance extraction quality and speed. The ELUSNet classification model was employed to handle the complexity of HSI feature extraction and classification. Moreover, this model successfully handled the problem of the ReLu and BN by using ELU AF. The ELUSNet outperformed the other five well-known models by more than 11% in terms of accuracy and training time. Future work will focus on improving feature extraction reduction methods and testing speed.

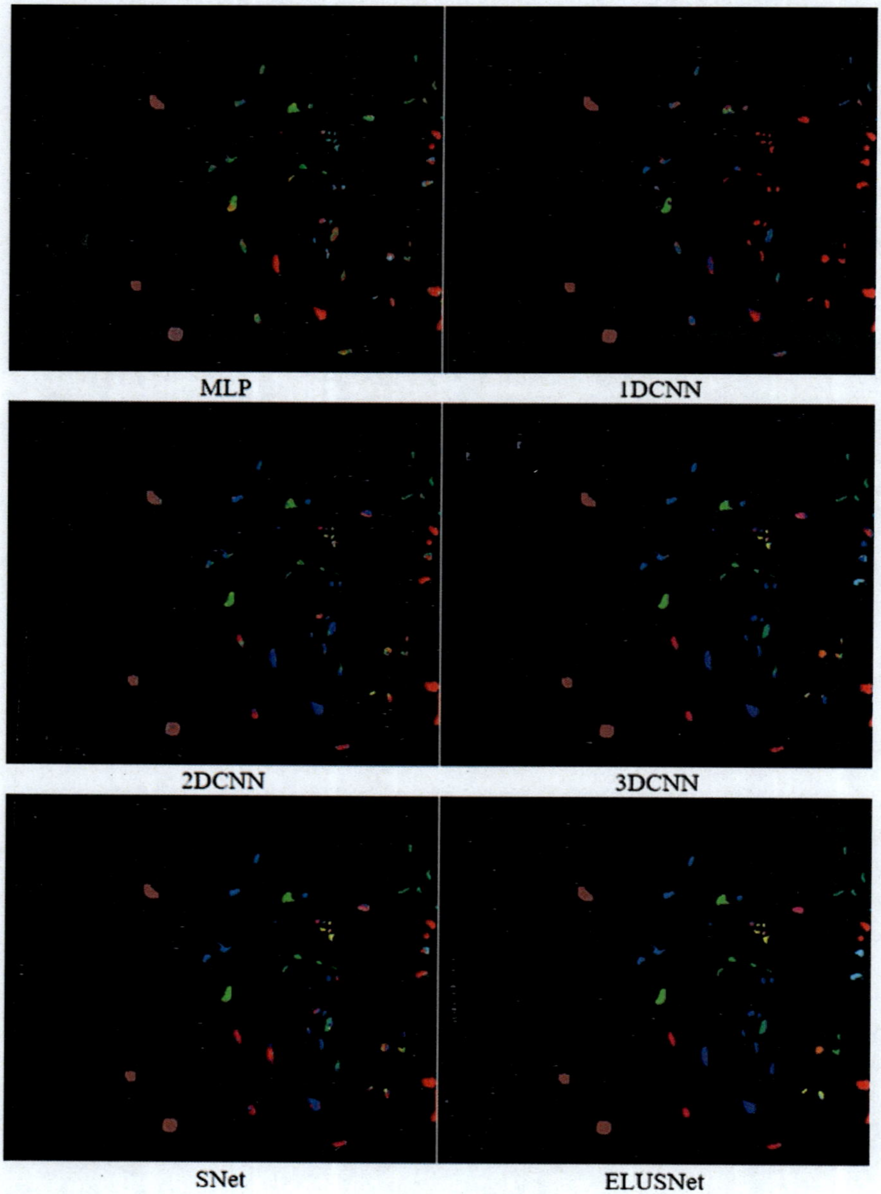

Fig. 5 The prediction of the six models for the KSC dataset

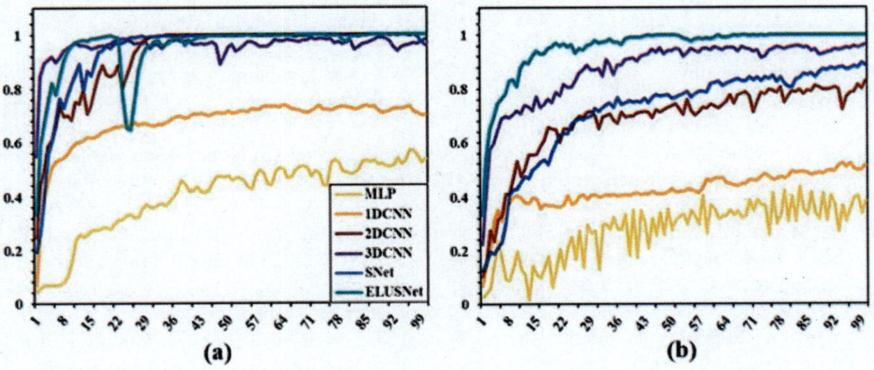

Fig. 6 The training accuracy for all models in each epoch

Funding This work was supported by the National Natural Science Foundation of China (Grant No. 62150410434) and partly by LIESMARS Special Research Funding.

References

1. Y. Gu, J. Chanussot, X. Jia, J.A. Benediktsson, Multiple kernel learning for hyperspectral image classification: a review. IEEE Trans. Geosci. Remote Sens. **55**, 6547–6565 (2017). https://doi.org/10.1109/TGRS.2017.2729882
2. M. Imani, H. Ghassemian, An overview on spectral and spatial information fusion for hyperspectral image classification: current trends and challenges. Inf. Fusion. **59**, 59–83 (2020). https://doi.org/10.1016/j.inffus.2020.01.007
3. Y. Xu, B. Du, F. Zhang, L. Zhang, Hyperspectral image classification via a random patches network. ISPRS J. Photogramm. Remote Sens. **142**, 344–357 (2018). https://doi.org/10.1016/j.isprsjprs.2018.05.014
4. J. Zhu, L. Fang, P. Ghamisi, Deformable convolutional neural networks for hyperspectral image classification. IEEE Geosci. Remote Sens. Lett. **15**, 1254–1258 (2018). https://doi.org/10.1109/LGRS.2018.2830403
5. N. He, M.E. Paoletti, J.M. Haut, L. Fang, S. Li, A. Plaza, J. Plaza, Feature extraction with multiscale covariance maps for hyperspectral image classification. IEEE Trans. Geosci. Remote Sens. **57**, 755–769 (2019). https://doi.org/10.1109/TGRS.2018.2860464
6. S. Mei, J. Ji, Y. Geng, Z. Zhang, X. Li, Q. Du, Unsupervised spatial-spectral feature learning by 3D convolutional autoencoder for hyperspectral classification. IEEE Trans. Geosci. Remote Sens. **57**, 6808–6820 (2019). https://doi.org/10.1109/TGRS.2019.2908756
7. A. Sellami, M. Farah, I. Riadh Farah, B. Solaiman, Hyperspectral imagery classification based on semi-supervised 3-D deep neural network and adaptive band selection, Expert Syst. Appl. **129**, 246–259 (2019). https://doi.org/10.1016/j.eswa.2019.04.006
8. W. Wang, S. Dou, S. Wang, Alternately updated spectral–spatial convolution network for the classification of hyperspectral images. Remote Sens. **11** (2019). https://doi.org/10.3390/rs11151794
9. C. Yu, R. Han, M. Song, C. Liu, C.-I. Chang, A simplified 2D-3D CNN architecture for hyperspectral image classification based on spatial–spectral fusion. IEEE J. Sel. Top. Appl.

Earth Obs. Remote Sens. **13**, 2485–2501 (2020). https://doi.org/10.1109/JSTARS.2020.2983224

10. M. Bandyopadhyay, Multi-stack hybrid CNN with non-monotonic activation functions for hyperspectral satellite image classification. Neural Comput. Appl. **33**, 14809–14822 (2021). https://doi.org/10.1007/s00521-021-06120-5
11. D. AL-Alimi, M.A.A. Al-qaness, Z. Cai, A. Dahou, Y. Shao, S. Issaka, Meta-learner hybrid models to classify hyperspectral images. Remote Sens. **14**, 1038 (2022). https://doi.org/10.3390/rs14041038
12. A. Mohan, M. Venkatesan, HybridCNN based hyperspectral image classification using multi-scale spatiospectral features. Infrared Phys. Technol. **108**, 103326 (2020). https://doi.org/10.1016/j.infrared.2020.103326
13. M.E. Paoletti, J.M. Haut, J. Plaza, A. Plaza, Deep learning classifiers for hyperspectral imaging: a review. ISPRS J. Photogramm. Remote Sens. **158**, 279–317 (2019). https://doi.org/10.1016/j.isprsjprs.2019.09.006
14. S.K. Roy, G. Krishna, S.R. Dubey, B.B. Chaudhuri, HybridSN: exploring 3-D–2-D CNN feature hierarchy for hyperspectral image classification. IEEE Geosci. Remote Sens. Lett. **17**, 277–281 (2020). https://doi.org/10.1109/LGRS.2019.2918719
15. S.U. Amin, M. Alsulaiman, G. Muhammad, M.A. Mekhtiche, M. Shamim Hossain, Deep learning for EEG motor imagery classification based on multi-layer CNNs feature fusion. Futur. Gener. Comput. Syst. **101**, 542–554 (2019). https://doi.org/10.1016/j.future.2019.06.027
16. Z. Sun, L. Xie, D. Hu, Y. Ying, An artificial neural network model for accurate and efficient optical property mapping from spatial-frequency domain images. Comput. Electron. Agric. **188**, 106340 (2021). https://doi.org/10.1016/j.compag.2021.106340
17. N. Wu, S. Weng, J. Chen, Q. Xiao, C. Zhang, Y. He, Deep convolution neural network with weighted loss to detect rice seeds vigor based on hyperspectral imaging under the sample-imbalanced condition. Comput. Electron. Agric. **196**, 106850 (2022). https://doi.org/10.1016/j.compag.2022.106850
18. L. Zhang, D. An, Y. Wei, J. Liu, J. Wu, Prediction of oil content in single maize kernel based on hyperspectral imaging and attention convolution neural network. Food Chem. 133563 (2022). https://doi.org/10.1016/j.foodchem.2022.133563

A Healthcare System Based on Fog Computing

Maha Abdulaziz Alhazzani, Samia Allaoua Chelloug, Reema Abdulaziz Alomari, Maha Khalid Alshammari, and Reem Shaya Alqahtani

Abstract Throughout the years, cloud computing has been a rapidly growing technology that offers different services for end-users. Despite the strengths of cloud computing, some limitations including the delay have been indicated by developers and researchers. Therefore, fog computing has gained an increasing interest due to its efficiency and effectiveness in terms of providing storage and processing resources at the proximity of end devices. This paper investigates the concept of fog computing to analyze some medical data that represents the measurement of vital signs of the human body. The measurements are collected through dedicated sensors that can measure body temperature, oxygen saturation and pulse rate. Collected data is sent to the suitable device (cloud or fog) for processing and analysis. More specifically, we propose a system that supports patients and nurses by providing two mobile applications. The fog node uses the nurses' application which is designed mainly to assist nurses with appointments management, data analysis, data processing, communication with patients and their relatives. Whereas patient's application is developed to receive a fast-paced data analysis of the vital signs. The main features of the proposed mobile applications are delay minimization, security, authentication, usability and scalability.

Keywords Fog computing · Cloud computing · WSN · Mobile applications · Vital signs · Healthcare · E-Health · Sensors

1 Introduction

Cloud computing has emerged recently to support many IoT applications, including healthcare applications. It has provided lots of computing services to the end users by

---

M. A. Alhazzani (✉) · S. A. Chelloug (✉) · R. A. Alomari · M. K. Alshammari · R. S. Alqahtani
Department of Information Technology, College of Computer and Information Sciences, Princess Nourah Bint Abdulrahman University, Riyadh, Saudi Arabia
e-mail: mahazzanii@gmail.com

S. A. Chelloug
e-mail: sachelloug@pnu.edu.sa

M. Abd Elaziz et al. (eds.), *International Conference on Artificial Intelligence Science and Applications (CAISA)*, Advances in Intelligent Systems and Computing 1441,
https://doi.org/10.1007/978-3-031-28106-8_5

connecting a huge number of IoT devices together [1]. However, majority of these devices have limitation that must be observed such as being resource-constrained [1]. Moreover, the efficiency of cloud computing solutions is related to its ability to analyze data while ensuring scalability and reliability. Furthermore, most IoT healthcare applications that depend on cloud computing are impacted by the latency issue as cloud datacenters are centralized and multi-hop communication is adopted between IoT devices and the cloud. Thus, the fog computing concept gained increasing attention due to its efficiency and effectiveness in transferring data quickly and safely. Fog computing extends the cloud computing to fully support IoT applications in a sufficient manner [2]. In addition, fog computing proved to have more advanced processing and closer storing mechanisms at the edge of the network. Hence, fog computing can be referred to as Edge computing. The two terms are used interchangeably [2]. So far, this paper investigates fog computing concept for analyzing some medical data representing measuring vital signs of the human body. Our system has the following features:

- It provides two mobile applications to support communications between the patients and the nurses.
- It allows to measure the most important vital signs for patient and send it to be analyzed.
- It decides on the best location (fog or cloud) for analyzing medical data.
- It aims at minimizing the latency of IoT devices.

In this paper, we propose a system composed of a set of sensors for measuring body temperature, heart rate and oxygen Saturation. In addition, the system includes a set of fog nodes that construct the fog layer. Also, two mobile applications for patients and nurses are proposed. The first application is the patient mobile application which allows the patient to book an appointment, select the suitable fog node, wearing the sensors, sending his/her collected data (vital signs) from sensors to the appropriate fog node/cloud. The system provides communication between the patient and nurse the application. Patient can view his/her medical reports which are stored in the cloud. The role of the second mobile application which is nurse mobile application is to enable the nurse to view patients, appointments, the medical report of any patient, find the relative information for patients and communicate with the patient. The most important part in nurse mobile application is the processing of the data that are collected from patient's sensor and analyzing them.

Thus, this paper is organized as follows. First, Sect. 2 introduces the related works. Then, Sect. 3 explains the architecture of our proposed system. Next, Sect. 4 illustrates and discusses our implementation details and results. Finally, Sect. 5 concludes this paper and discusses potential future work.

2 Related Works

Silva et al. [3] suggest integrating fog computing with blockchain. Fog Computing techniques are mainly used to provide availability and performance, whereas Blockchain-based strategies are used to ensure the privacy needed for any medical domain. This paper has proposed an architecture including four layers for medical records management. Namely, the cloud, fog, application, and sensor layers are available. The role of the fog layer is to manage a set of medical records and validate the access to the data through blockchain technology. Each fog node function is represented as a blockchain miner and provides a REST API. The developed mobile application allows also to request and grant data access. In addition, some of the non-functional requirements have been tested including the privacy and interoperability.

Ben Hassen et al. [4] suggest a home hospitalization system which integrate techniques based on Fog computing, Cloud computing and the IoT. The fundamental feature of the system is to allow patients to receive treatments and recover in their own homes, where the health of the patients and the environmental factors of their hospitalized rooms are monitored periodically, by using vital signs sensing unit and environmental sensing units that are equipped in the hospitalization rooms and provided mobile applications developed for this purpose. The architecture of the proposed system is composed of two sections, the proposed system's architecture and the proposed cloud computing architecture. The proposed system's architecture suggests that there are three interconnected parts, the system application that users will use (patient, relative, nurse, doctor, and administration) and the cloud computing NoSQL database Cloud computing to hold the REST API and data storage where the Microsoft Azure Cloud Platform will be used. Furthermore, they have relied on Azure Cosmos DB. that will be connected to the web server REST API which will be processed by the hospitalization room which is concerned with the environmental sensing unit which are a set of sensors that will measure the signs of the patient's surrounding environment with a mobile application that will play the role of the fog node and will allow the measurements of the environmental factors of the hospitalization room in real time, whereas the patient's medical data is observed through a vital signs sensing unit with mobile application that plays the role of the fog node. The proposed system in this paper allows doctors, patients and everyone who has an interest in following up and managing this process through their mobile devices. The proposed system in this paper has its non-functional features which are low cost, reliability, and security in addition to its ability to solve the problems that hospitals are currently experiencing, it can significantly reduce the burden on them. This system has been well accepted by patients and doctors alike according to the usability assessment results.

Oueida et al. [5] have proposed an RPN framework in 2 categories (patient related and process related) that uses Petri net technology integrated with both edge computing and custom cloud suitable for ED systems. The designed framework is used to model non-consumable resources and is described and validated by developing a wireless single chip to collect and monitor real time data and heart activities and enhance it by analyzing these collected data into researches for process and policies in health sector. The proposed framework allows to assign resources in the edge, while resource scheduling is performed in the cloud. RPN is suitable for a real-life scenario where patient waiting time and resource utilization are the performance indicators to be optimized and modeled. The simulation of the framework highlights the improvements in LoS patient waiting time and resource utilization.

Elgendy et al. [6] propose a reliable remote health care system for patients, especially in nursing homes during the Corona virus period, and it is called RHHM, where patients can be supervised and their health condition diagnosed while they are in their homes, and the system exploits the benefits of fog layers with high services The level such as storage and data processing, in addition to the presence of a camera along the wireless sensors for more reliability, accuracy and efficiency, the system diagnoses the patient's condition while the camera relies on taking a picture of the patient to analyze his/her emotions and body. But some diseases require detection and not relying on sensors, such as internal medicine, toothache, muscle pain and headache. The technologies used in this system are the IoT, cloud computing and fog computing.

Kharel et al. [7] proposed a smart health monitoring system that integrates Long Range (LoRa) wireless communication and low power wireless sensor network (WSN). Long Range (LoRa) radio is the connectivity solution used by the authors to allow IoT devices to communicate over long range communication while consuming low energy. The architecture of the proposed system is based on hierarchical FC, where the main fog nodes are the health centers and the secondary fog servers are the hospitals. The health centers' fog nodes are the edge user devices whereas health centers can be regarded fog nodes for hospitals. Edge users are equipped with sensors, WBS, or other wearable medical devices. The LoRa gateway is used to convert signals produces from the medical devices used the end users. Then, the information is sent to the health centers to be analyzed. Since LoRa features long range transmission, the sensor's produced data can be sent with the absence of the internet to a health center located many kilometers away. Also, health centers store and back up the data received from edge users based on the medical conditions of each case. The backed-up data can be viewed in the actual time. Moreover, authors evaluated the proposed system by using a testbed for the architecture proposed and assessed the performance of the network. The results founded by test had shown efficient performance of LoRa beside FC. The suggested system helped to obtain a monitoring system of user's biometric information remotely and tens of kilometers away. Also, it helped to minimize burden on the cloud as well as having an energy-efficient health monitoring system.

Farahani et al. [8] They discuss the possibility of applying the concept IoT in medicine and healthcare system by providing a comprehensive architecture for the IoT e-health ecosystem and relying on Fog computing to be able to reduces the time for analyzing and processing circumstances and time-sensitive data such as Myocardial Infarction (MI).

Paper [9], proposed an electronic architecture for health monitoring that relies on sensors and relies on storing and handling patients' data on cloud and fog computing. The main objective of this research is to determine problems in the suggested architecture in addition to suggesting strategies to reduce application downtime, due to the negative consequences of application failure and service unavailability even for a few seconds, such as inconvenience or misdiagnosis and may lead to death. In this paper, random models were proposed using the SPN and RBD approach, and a prototype was developed and implemented with different types of fog devices in four different geographical locations, and the results were different in terms of service time and performance, so before deciding the best architecture, it is necessary to prioritize system requirements first, as each device has some specific feature.

Klonoff [10] proposed an Internet-based computing architecture that relies on fog computing to store and analyze medical data generated from a number of IoT devices remotely by distributed servers. The wireless devices used by end users are limited to diabetes devices which can be concluded by insulin pens or pumps, continuous or blood glucose monitors and closed-loop systems. As mentioned in this paper, the fog computing has promising potential over cloud computing in which it has achieved greater privacy and security, faster data transmission speed, greater control over foreign countries' data where laws may limit the access, less dependence on limited bandwidth, lower costs due to local data derived from sensors.

Paper [11], proposed a heterogeneous cloud assisted communication framework and consist of fog, cloud, data collection and application layers to treat separately the non-real time and real-time data. Packet delivery ratio and throughput are taken into account to improve QoS. The proposed solution performs the data scheduling to different layers for data preprocessing and data storage so that decision can be taken. For this, APSO has been considered which schedules the data traffic to different devices of fog and cloud layers so that later, different machine learning algorithms can be applied for further processing. The results obtained clearly indicate the significance of the algorithm.

Considering the challenges presented in the current solutions, we have proposed a solution to improve the efficiency of healthcare applications by sitting two mobile applications one for the patient and one for the nurse while investigating the concept of a fog node that combines sensors to process and analyze a patient's vital signa, both remotely and efficiently. In addition to protecting patients' data through implementing the AES algorithm and the authentication using two methods, either by email or phone number. Moreover, compared to other similar systems the proposed system provide a measurement of three vital signs which are: body temperature, heart rate, and oxygen saturation.

3 System Architecture

Figure 1 shows the layered view of proposed our system architecture. Layered architecture is the simplest form of the software's architectural pattern. The components of

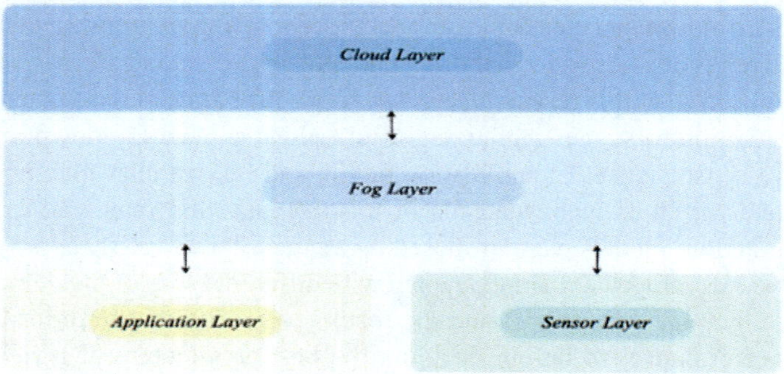

Fig. 1 Proposed layered architecture

the system are organized in horizontal layers, where all the layers are interconnected but not dependent on each other. It is easy to test the components belong to specific layer separately. In addition, it is simple and easy to implement. 4 layers were used, and they are as follows [3]:

- Sensor Layer: It consists of one sensor device and it responsible for monitoring the vital signs of patient [3].
- Application Layer: This layer enables interaction with the system and the ability to access and manipulate data [3].
- Fog Layer: It is responsible for receiving data generated by the devices in the application layer and the sensors, and managing and storing a subset of the data for processing [3].
- Cloud Layer: It contains a database of all patients' records that are stored permanently [3].

3.1 Patient's Mobile Application

The mobile application of the patient will open the patient allow him/her to book or view an appointment. In addition, to reschedule it or cancel it. The mobile application enables also the patient to check the timing of his/her appointment. If the appointment time has started, the mobile application of the patient executes the steps explained in Fig. 2.

Since fog nodes are considered limited-resources and cannot handle all requests [12], tasks may experience some type of delay. Computational offloading is one of the solutions used to minimize the latency of tasks and improve quality of service (QoS) of user applications [12]. Computational offloading means sharing the workload among the nodes in the same layer or with the upper layers in order to improve the performance [12].

Patient's mobile application is responsible for finding suitable fog node to send their data to. Each sensor and fog node will be having one parameters of interest,

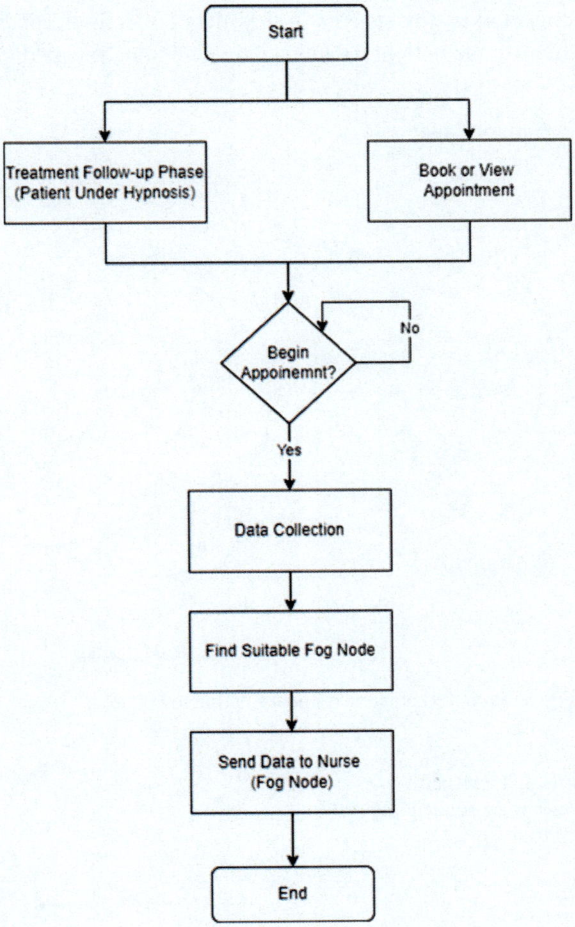

Fig. 2 Flowchart of the mobile application of the patient

which is RAM. In order for the patient's application to find a suitable fog node, each fog node will send its available RAM to the patient's application. Patient's application will decide whether this fog node is suitable or not. Figure 3 shows the interactions between cloud layer, sensor layer and the fog layer with the essential arguments for each layer. A fog node is considered suitable if and only if the RAM of the fog node is larger than or equals the capacity needed for the sensor. In our mobile application, patients are able to pick a fog node from a list of available fog nodes. In case of an unavailability of all fog nodes, patient's mobile application will send its data to the cloud until one of the fog nodes is available and suitable. Figure 4 illustrates the flow of 'Find a Suitable fog node' process. When the patient sends his/her data to the suitable fog node, the nurse's fog node will process the data and encrypt it using AES algorithm, and finally the cloud will store the data after being encrypted and processed. After the process of searching and selecting the suitable fog node, the patient mobile application will send all the collected data that were previously collected by the sensors in the "data collection process" to the nurse (fog node) to analyze the patient's data and create/update the medical report of the patient.

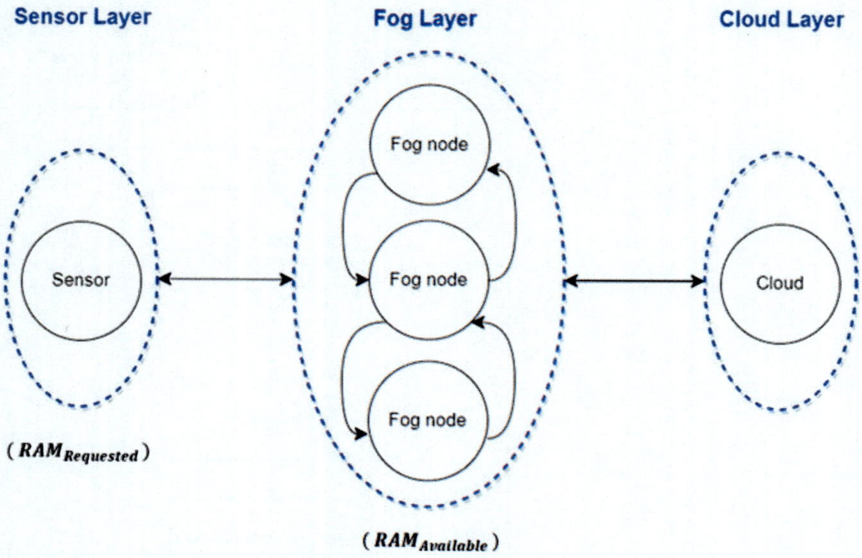

Fig. 3 Proposed process to find a suitable fog node

Fig. 4 Flowchart for selecting a suitable fog node

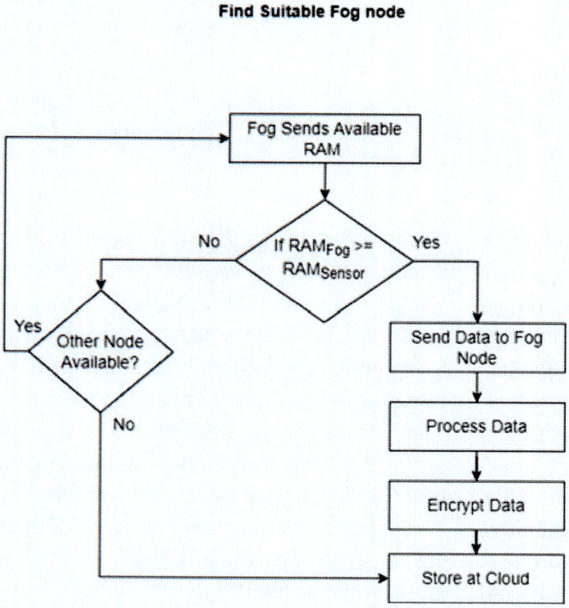

3.2 Nurse's Mobile Application

A nurse shall open her mobile application to view all the scheduled appointments as shown in Fig. 5. The appointments have specific date, time and information about the patient linked to it. The nurse must always keep up with the appointment timing to see when it is started and begin the appointment with the patient. In the patient side, the patient mobile application will perform an algorithm of deciding whether the fog node is suitable or not. This method will result in having either multiple suitable fog nodes or none at all. In case of considering a fog node a suitable, the patient will have the ability to pick this fog node or any other suitable fog nodes. In this process, nurses will know whether their fog nodes are suitable or not. If nurse's fog node is suitable, nurse shall wait if the patient will pick her as a fog node or not. Otherwise, if nurse's fog node is not suitable patient cannot pick the fog node and nurse can leave the appointment. After ensuring that fog nodes are suitable, nurses shall wait to receive an indicator from the patient to acknowledge whether their fog nodes are picked by the patient picked or not. If nurse's fog node is picked by the patient, nurse shall prepare to receive patient data and perform the processing. Elsewhere, the nurse can leave the appointment since patient did not pick their fog node.

3.3 Data Analysis

The data received are gathered from sensors, which had computed: body temperature, pulse rate and SPO2. The data will be labelled according to the well-known medical measurements of vital rates, to classify patients into normal state or critical state.

- A case is considered normal, if the vital rates of the patient are correspondent to the measurements of the normal vital rates as shown in Table 1.
- A case is considered critical, if the vital rates of the patient are correspondent to the measurements of the critical vital rates as the shown in Table 1 [13–16].

4 Implementation Details

The implementation of the e-healthcare system is done by developing two applications in Android Studio by using Java language. One application is dedicated for the patients and one is for the nurses. Firebase NoSQL database is used along with the applications to authenticate users and to store users' data. The hardware setup is concluded by using Arduino Uno board to connect the medical sensor MAX30102 using Male to Male wires. Data generated by MAX30102 is preprocessed using Arduino IDE. Then, data will be received in Visual Studio Code using Python language to send data to the available fog node.

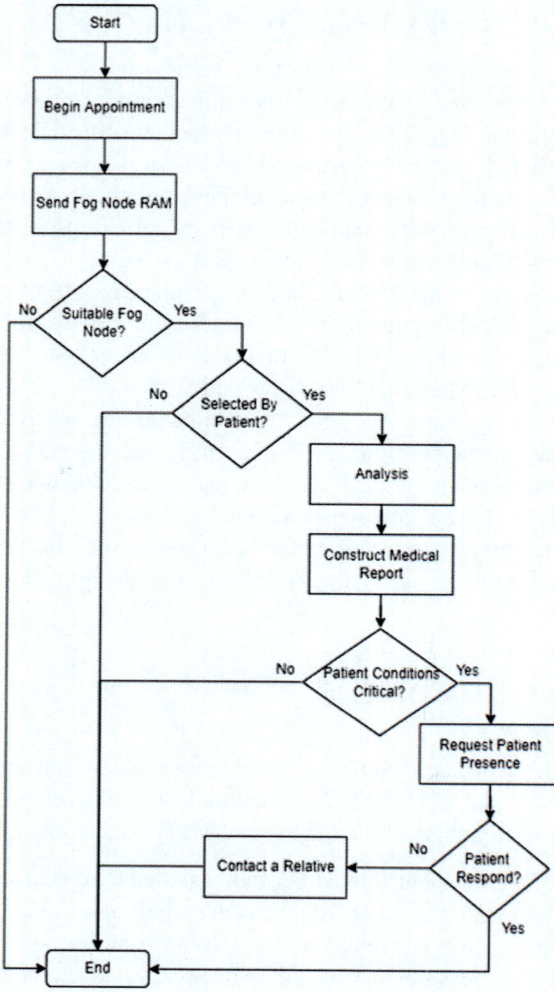

Fig. 5 Flowchart for the nurse's mobile application

Table 1 Measurements of the critical vital rates

Test case	Conditions
Normal	36.5 °C≤ Body temperature ≤ 37.2 °C
	60 BPM ≤ Pulse rate ≤ 100 BPM
	95% ≤ Oxygen Saturation ≤ 100%
Critical	35.0 °C < Body Temperature < 36.5 °C 37.2 °C < Body Temperature < 41.0 °C
	43 BPM < Pulse rate < 60 BPM 100 BPM < Pulse rate < 200 BPM
	67% < Oxygen Saturation < 95%

Patient application is responsible of selecting the suitable fog node as illustrated in Fig. 6. The vital signs data generated at the patient side needs a memory to process. Thus, the patient application will check each RAM the fog nodes has sent and single out one suitable fog node. Then, it will send the selection to Firebase to allow nurse application to acknowledge that the fog node has been chosen. The interface of RAM information is shown in Fig. 7. In other cases, fog nodes might not be suitable where the capacity of the fog node is not enough to process data or no fog nodes are available at all. In these cases, the cloud layer is used to support the fog layer by storing the data. After the patient chooses the suitable fog node, he must wear the sensor to calculate his vital signs as shown Fig. 8, and after collecting the readings, it will be sent from Arduino IDE to Visual Studio Code, where it will be saved in a CSV file for preprocessing the data and ignore the abnormal readings to make sure that the file contains at least one row that contains somewhat normal readings/values. If the condition is met, file will be uploaded to the Storage Firebase for data processing step, if not, a new reading will be received from the Arduino IDE.

In data processing, the nurse application receives all sensor readings from the Firebase repository. Data cleaning process starts by eliminating out of boundary values, relying on Table 1, both normal and critical ranges of the three vital signs are

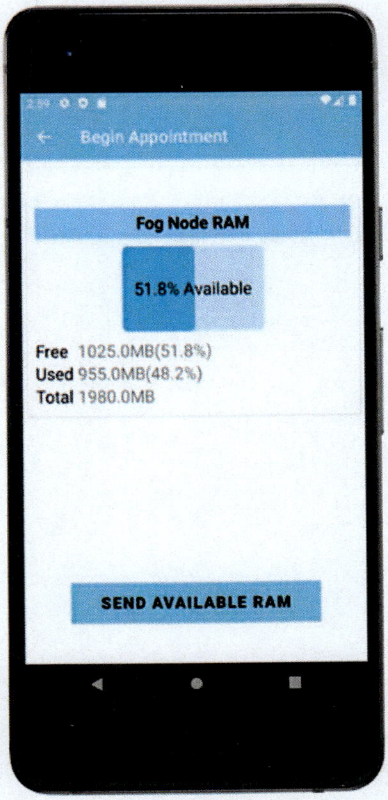

Fig. 6 Interface for finding a suitable fog node

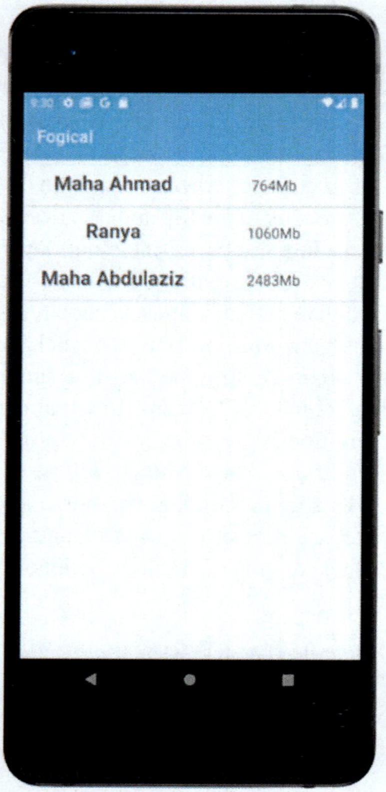

Fig. 7 Interface for RAM information

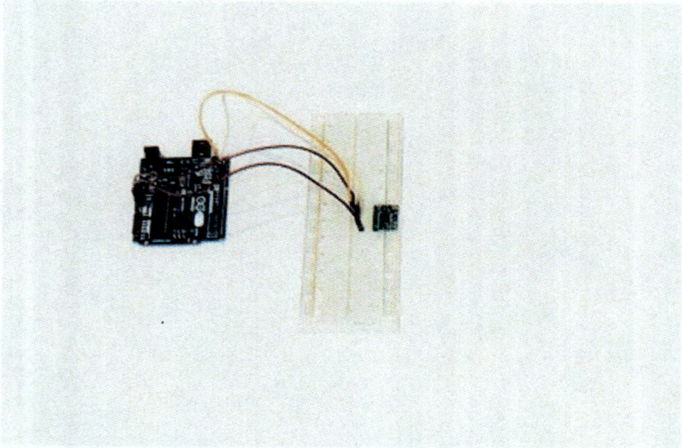

Fig. 8 Arduino circuit

listed. Moreover, Auto-diagnosis service is provided to increase the QoS features for the applications. The sensor used generate multiple lines of readings which can be impacted by how patients wear the sensor and other environmental factors. To get the most accurate value, these readings are averaged. Finally, patient medical data

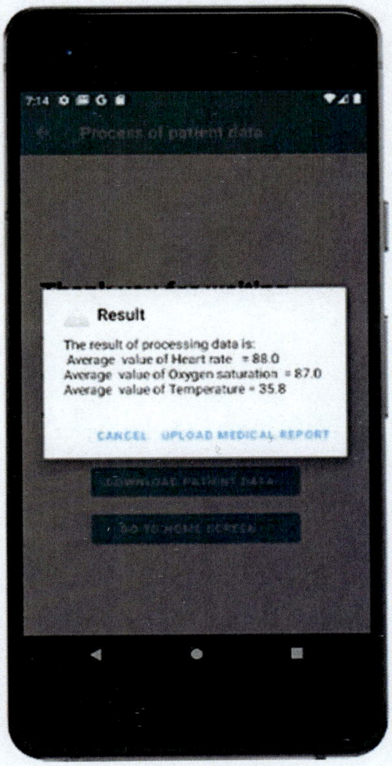

Fig. 9 The result of processed data

is protected by implementing the AES algorithm. Figure 9 illustrates the result of processed data. After the data processing is finished, the encrypted data is stored in the database and can be retrieved and displayed as a medical report for both nurses and the desired patient in the application when needed.

The communication between the nurse and patient is built using firebase. Using identifiers for both the nurse and the patient a communication channel was built for them. In the nurse application, a list of the patients with their information and the option to communicate with them as illustrated in Figs. 10, 11 and 12. The nurse is the one who can start the conversation for the first time, because the step of fog node implementation must be done first as it was explained previously. Whereas in the patient application, a list of nurses who have previously checked the patient's vital signs, and previously communicated with them is present and retrieved from the database as a list of conversations designed to make the patient recommunicate with them when needed.

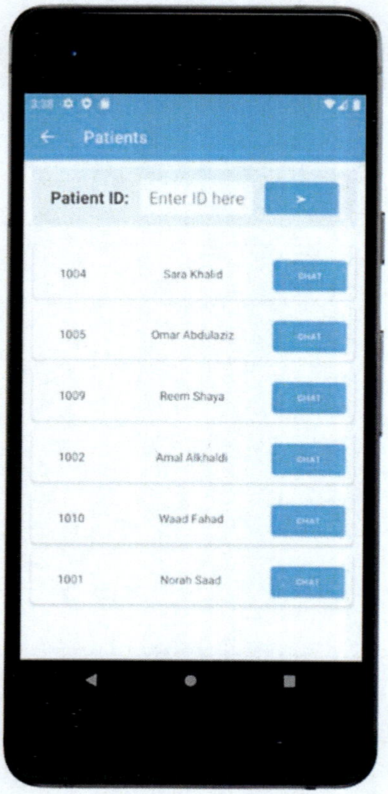

Fig. 10 Patients' list

5 Conclusion and Future Work

This paper proposes the development of a healthcare system based on fog computing that overcomes various performance issues especially the delay that was faced in cloud computing. In this paper, the concept of fog computing which imposes the processing and analyzing data obtained from sensors in the fog database is implemented. As the proposed system includes two main databases, one of them is the fog database that will store the RAM information as well as the vital signs data of patients who have met the basic algorithm requirement "find the suitable fog node" where the RAM value between the nurse device and the sensor is suitable. While the other database belongs to the cloud and stores the cases that did not meet the algorithm requirement. The proposed system publishes a set of available fog nodes at the same appointment time for the patient, and through its application, the patient will be enabled to the selection of the appropriate fog node according to that algorithm. The system serves patients and nurses to analyze vital signs and three main measurements namely body temperature, heart rate and oxygen saturation through two mobile applications for nurse and patient each with different functions explained earlier. The system was supported by the application of a security algorithm that has

Fig. 11 Chat list

a short response time to protect patients' data, authentication of users using two methods; E-mail and phone number, building a chatting sub-application to allow communication between the patient and the nurse, managing appointments for both the nurse and the patient, the automatic diagnosis feature, which the nurse's application has been programmed on the normal and abnormal values of vital signs so that it facilitates and leads to raising the efficiency of healthcare applications, and it is considered an application for one of the most important objectives of the proposed research, which will eventually lead to an increase in the usability and a reduction in delays. However, the proposed system is limited to using a single sensor to measure patients' vital signs, as the sensor itself has limitations in terms of distance so the sensor must be close to the mobile device for the data to arrive. In addition, some of the data produced by the sensors is not as precise as the newer devices used to measure human vital signs. As a future work, we suggest using wearable devices as an alternative and building a machine learning model to predict patient data and train it to do the processing and prediction. In terms of communication between the patient, the nurse and the system in general developing a chatbot will certainly be effective.

Fig. 12 Chat example

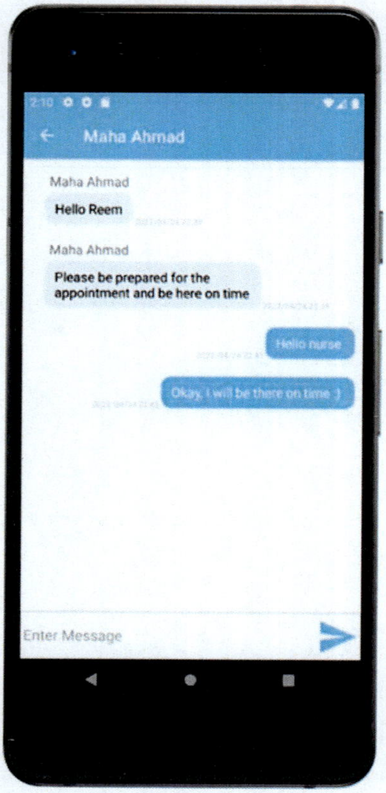

Conflicts of Interest The authors declare that they have no conflicts of interest to report regarding the present study.

References

1. A. Rahmani, P. Liljeberg, J. Preden, A. Jantsch, *Fog Computing in the Internet of Things* (Springer, Intelligence at the Edge. Cham, 2018)
2. Z. Mahmood, M. Ramachandran, Fog computing: concepts, principles and related paradigms, in *Fog Computing: Concepts, Frameworks and Technologies* (Springer, 2018), pp. 3–21
3. C. Silva, G. Aquino, S. Melo, D. Egídio, A fog computing-based architecture for medical records management. Wirel. Commun. Mob. Comput. **2019**, 1–16 (2019)
4. H. Ben Hassen, N. Ayari, B. Hamdi, A home hospitalization system based on the Internet of things, Fog computing and cloud computing. Inf. Med. Unlocked **20**, 100368 (2020)
5. S. Oueida, Y. Kotb, M. Aloqaily, Y. Jararweh, T. Baker, An edge computing based smart healthcare framework for resource management. Sensors **18**(12), 4307 (2018)
6. F. Elgendy, A. Sarhan, M. Alshewimy, Fog-based remote in-home health monitoring framework. Int. J. Adv. Comput. Sci. Appl. **12**(6), 247–254 (2021)
7. J. Kharel, H. Reda, S. Shin, Fog computing-based smart health monitoring system deploying LoRa wireless communication. IETE Tech. Rev. **36**(1), 69–82 (2018)

8. B. Farahani, F. Firouzi, V. Chang, M. Badaroglu, N. Constant et al., Towards fog-driven IOT ehealth: promises and challenges of IOT in medicine and healthcare. Futur. Gener. Comput. Syst. **78**, 659–676 (2017)
9. G. Santos, P. Takako Endo, M. Ferreira da Silva Lisboa Tigre et al., Analyzing the availability and performance of an e-health system integrated with EDGE, fog and cloud infrastructures. J. Cloud Comput. **7**(16) (2018)
10. D.C. Klonoff, Fog computing and edge computing architectures for processing data from diabetes devices connected to the medical internet of things. J. Diabetes Sci. Technol. **11**(4), 647–652 (2017)
11. R. Chudhary, S. Sharma, Fog-cloud assisted framework for heterogeneous internet of healthcare things. Procedia Comput. Sci. **184**, 194–201 (2021)
12. F. Alenizi, O. Rana, *Minimizing Delay and Energy in Online Dynamic Fog Systems* (2020), pp. 139–158
13. National library of Medicine. https://medlineplus.gov/ency/article/001982.htm
14. Mayo clinic. https://www.mayoclinic.org/healthy-lifestyle/fitness/expert-answers/heart-rate/faq-20057979
15. B.K. Peterson, *Physical Rehabilitation, Evidence-Based Examination, Evaluation, and Intervention* (2007), pp. 598–624
16. M.K. Park, *Park's Pediatric Cardiology for Practitioners*, 6th edn. (2014), pp. 108–116

An Improved Arithmetic Optimization Algorithm with Differential Evolution and Chaotic Local Search

Aminu Onimisi Abdulsalami, Mohamed Abd Elaziz, Yousif A. Al Haj, and Shengwu Xiong

Abstract These days the notion to hybridize metaheuristics has become popular among researchers due to the advantages it presents to tackle complex and difficult optimization problems. This method makes it easy to combine the strength of two or more algorithms to better traverse a search space. The arithmetic optimization algorithm (AOA) is a recently developed metaheuristic that has received enormous attention among researchers due to its simple and robust nature. The algorithm applies selective pressure using the arithmetic operators (+, −, ×, ÷) to explore and exploit a search space. This action is highly desirable when the process has identified the optimal region, but could be detrimental in situations where the algorithm needs to visit new and diverse regions to escape getting trapped. To address this problem, this study combines AOA and a modified differential evolution (DE) mutation strategy to help solutions jump out of local optimum and prevent premature convergence. Furthermore, we added a chaotic local search operator into AOADE to help improve the quality of the best solution at the end of each iteration. Experimental results on the IEEE CEC 2014 benchmark functions indicate that our proposed method obtained

A. O. Abdulsalami · S. Xiong (✉)
School of Computer Science and Technology, Wuhan University of Technology, Wuhan 430070, China
e-mail: xiongsw@whut.edu.cn

A. O. Abdulsalami
e-mail: aoabdulsalami@whut.edu.cn

M. A. Elaziz
Department of Mathematics, Faculty of Science, Zagazig University, Zagazig 44519, Egypt

Faculty of Computer Science and Engineering, Galala University, Suze 435611, Egypt

Artificial Intelligence Research Center (AIRC), Ajman University, 346 Ajman, United Arab Emirates

A. O. Abdulsalami
Department of Computer Science, Ahmadu Bello University, Zaria, Nigeria

Y. A. Al Haj
Sanaa Community College, 5695 Sanaa, Yemen

M. Abd Elaziz et al. (eds.), *International Conference on Artificial Intelligence Science and Applications (CAISA)*, Advances in Intelligent Systems and Computing 1441,
https://doi.org/10.1007/978-3-031-28106-8_6

better results in at least 20 out of the 30 benchmark functions tested as compared to other state of the art algorithms.

Keywords Arithmetic optimization algorithm (AOA) · Differential evolution (DE) · Chaotic local search (CLS) · Exploration and exploitation

1 Introduction

Complex and difficult problems with many feasible solutions are known to exist in many areas of human endeavour. In the past decades, researchers and industry practitioners have kept their focus on developing powerful and reliable solvers rooted in ideas from mathematical optimization to solve these problems. Optimization is simply the process of choosing the best element, according to some well-defined criteria, from a collection of available alternatives. It has become a much sought-after and game-changing method that has facilitated breakthroughs in areas like engineering design [1, 2], healthcare [3, 4], computer vision [5, 6], machine learning [7–11], transportation and logistics [12, 13], planning and scheduling [14, 15], and so on.

In general, finding the optimal solution is sometimes hard to control in optimization problems of scientific and industrial value, therefore, experts are usually comfortable with "near optimal solutions" otherwise referred to as "good solutions". Along this line, optimization algorithms can be classified into exact and approximate methods. The exact group hosts a number of algorithms that guarantee finding the optimal solution, but are not suitable for problems with complex structures. Also, note that the size of the problem is not always an indicator of how difficult the problem could be. Some well-studied exact algorithms are dynamic programming, constraint programming, branch and bound methods, and A* group of search techniques. In most cases, these algorithms work well with small instances of complex and difficult problems such as the quadratic assignment problem (30 objects), graph colouring problem (100 nodes) and the capacitated vehicle routing problem, just to mention some examples. The algorithms in the approximate grouped can find the optimal solution where the need arises, but there are no guarantees, and should be put to use where exact methods have failed. The approximate method can be further divided into specific heuristics and metaheuristics. The former is designed to adapt to the problem of interest, and by this, they take full advantage of the uniqueness of the problem. The latter is general purpose and does not take advantage of the peculiarities of a problem. In fact, they can be used as black box solvers to tackle any optimization problem, while providing good solutions in reasonable computational time.

They are known to implement some form of stochastic optimization which make them efficient solvers for global optimization problems. Metaheuristics can be further group into two classes: single solution based and population based algorithms. In the process of optimization the single solution based manipulates and transforms one solution at a time, while population based can manipulate and change a collection of solutions otherwise known as the population.

Another way to classify metaheuristics is to put them into the family of nature inspired or non-nature inspired algorithms. Many metaheuristics are inspired by things around us ranging from laws of evolution, social phenomenon, and animal foraging behaviour down to physical and chemical processes. For instance, the well-studied genetic algorithm (GA) [16], memetic algorithm (MA) [17] and differential evolution [18] are developed from ideas of biology; particle swarm optimization (PSO) [19], artificial bee colony (ABC) [20], cuckoo search (CS) [21], crow search (CSA) [22], Harris hawk optimization (HHO) [23] are inspired by swarm intelligence. Other metaheuristics inspired from non-natural phenomenon are equilibrium optimizer [24], black hole algorithm [25], sine cosine algorithm [26], gravitational search algorithm [27] and arithmetic optimization algorithm [28], all inspired by mathematics or physics. Here, it is worth noting that the development of these algorithms and their many variants is motivated by the No-Free-Lunch theory which states no single method can be powerful enough to solve all optimization problems. The study has two main motivations: the No-Free-Lunch theory and the need to tackle the weakness of the classical AOA which include local entrapment and premature convergence. However, during optimization it is expected that the whole population might converge to a local optima; this is desirable if the process has reached the region of the global best solution, otherwise it becomes counterproductive and requires an action to make the solutions visit new and diverse regions. To overcome this issue, we propose a new method called AOADE that combines AOA and a modified DE mutation scheme to improve the overall performance of the algorithm. Furthermore, we introduced the chaos theory to refine the fittest solution found in each iteration. This process has an exploitative effect on the population and help improve the quality of solutions. Finally, to assess the performance of our proposed method, we test the algorithm on the IEEE CEC 2104 benchmark suite [29], and compare our result against the results of the classical AOA and other state of the art metaheuristics. We selected AOA for this study due to simple and robust nature as well as it wide acceptability by researchers to tackle real world optimization problems.

The remaining part of the paper is organized as follows: the related works are discussed in Sect. 2. The mathematical foundation of AOA is described in Sect. 3. Mechanisms of the DE mutation operator and chaotic strategy used with AOA are described in Sect. 4. Section 5 highlights the experimental settings and present the results of experiments. Conclusions are drawn in Sect. 6.

2 Related Works

Regardless of being recently developed, many studies have put forward improved versions of the AOA and applied them to solve problems in diverse areas. For instance, the work of Li et al. [30], for the first time used the chaos theory with the AOA. The authors managed to separately embed ten chaotic maps into the main control parameters of AOA (Math Optimizer Accelerator and Math Optimizer Probability) to improve the overall performance of the algorithms. They tested the algorithm on the CEC2017 functions for global optimization and classical engineering design problems. In a similar work, the authors in [31] proposed a hybrid version of AOA, in which they combine the operators of GA with the operators of AOA to tackle the exploration defect of AOA, find high quality solutions and improve the convergence rate of the algorithm. They applied the algorithm to find subset of features of high dimensional data, so as to boost the classification accuracy of selected machine learning algorithms. From the experiments conducted, AOAGA outperformed state of the art methods in the feature selection tasks. In another hybrid version of proposed by Mahajan et al. [32], the authors combined the AOA and hunger games search (HGS) algorithm (AOA-HGS) to solve both and high and low dimensional problems. They reported superior results for AOA-HGS when compared to other state of the art metaheuristics. Abualigah et al. [33] proposed a hybrid version of AOA and Sine Cosine algorithm (SC). The SC operators were used to improve the exploitation ability of AOA. In addition, a further refinement is carried out on the final solution (best one) to improve its quality. The authors tested their method on ten benchmark functions and five engineering problems. Their method outperformed other state of the art algorithms tested. More recently Khodadadi et al. [34] proposed a dynamic version of the AOA (DAOA). Their algorithm introduce a dynamic candidate solution function, which replaced the math optimizer probability function of the classical AOA. The new function helps to improve the quality of solutions during the search process. They further applied DAOA to minimize the weights of four truss structures (37-bar planer truss, 72-bar space truss, 120-bar dome truss and 200-bar planar truss) under some frequency constraints, and the results obtained showed the superiority of DAOA over AOA and other state of the algorithms. Dahou et al. [35] presents a human activity recognition system based on two algorithms namely binary AOA and convolutional neural network (CNN). The latter learns and extract features from input data, while the former help select the most optimal features. In addition, the classical SVM is used to classify the selected features based on different activities. Ewees et al. [31] modified AOA to enhance its search strategies. The authors achieved this by using the classical GA operators in AOA. The proposed AOAGA was evaluated with several well-known benchmark datasets and a number of standard evaluation criteria were used.

3 Preliminaries

3.1 Arithmetic Optimization Algorithm

Mathematics is a subject of numbers, and the four well-known arithmetic symbols: addition (+), subtraction (−), multiplication (×) and division (÷) are among the oldest mathematical tools used to process numbers. These symbols have immensely contributed to the development of several sub areas of mathematics including mathematical optimization. The AOA proposed in 2021 by Abualigah et al. [28] is a population based metaheuristics inspired by the ancient mathematical operator (+, −, × and ÷). These operators are used to manipulate a population of individuals to solve optimization problems. In the beginning of the optimization, AOA generates a considerable amount of data to be used as the initial population using Eq. 1; and in this setting, a candidate solution is represented as a vector of real values.

$$X_{(i,j)} = lb_j + rand \times \left(ub_j - lb_j\right), i = 1, 2, \ldots, N \; and \; j = 1, 2, \ldots, D \quad (1)$$

where $X_{(i,j)} \in [lb_j, ub_j]$, and lb_j, ub_j are the lower and upper bound of the problem respectively; $rand$ draws a value from the uniform distribution in the range of 0, and 1. The next step is to apply the objective function to each solution in the population in order to obtain their fitness values. The process carries on to identify and label the solution with the most acceptable fitness before moving into the iterative phase. The search process of AOA is logically separated into two valuable operations in global optimization (exploration and exploitation), with the help of some control parameters: Math Optimizer Accelerated (MOA) and Math Optimizer Probability (MOP) functions. It is worth mentioning that, being able to strike a balance between these operations has enabled most successful metaheuristics. The MOA function is calculated according to Eq. 2.

$$MOA(t) = \text{Min} + t \times \left(\frac{\text{Max} - \text{Min}}{T}\right) \quad (2)$$

where t is the current iteration, which can keep a value in the range of 1 and the maximum number of iteration, (T); Min and Max are the highest and lowest values of the accelerated function.

3.1.1 Exploration Phase

In the exploration phase of AOA, the multiplication and division operators are solely responsible for driving the search process to visit new and diverse regions of search space. These operators produce highly dispersed values that makes it difficult for the process to reach its target without supporting operator like + or −. The authors of the AOA [28] have shown the influence of all four arithmetic operators on mathematical calculation using Fig. 1. Being able to search in this phase of the algorithm depends on the result of the MOA function given in Eq. 2; note that, if MOA $< r1$, (where $r1 \in U(0, 1)$), then exploration is performed, otherwise the algorithm jumps to the exploitation phase. The update equations that simulate the behaviour of the × and ÷ in this phase is given Eq. 3;

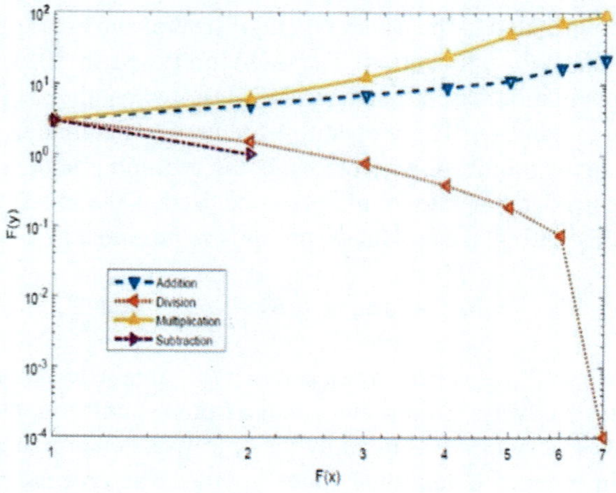

Fig. 1 The influence and behaviour of the arithmetic operators (+, −, ×, ÷) in two dimensions

$$X_{i,j}(t+1) = \begin{cases} best(x_j) \div (MOP + \varepsilon) \times ((ub_j - lb_j) \times \mu + lb_j) & r2 < 0.5 \\ best(x_j) \times (MOP + \varepsilon) \times ((ub_j - lb_j) \times \mu + lb_j) & otherwise \end{cases} \tag{3}$$

where $X_{i,j}(t+1)$ is the jth decision variable of the ith solution in next iteration, t is the current iteration, $best(x_j)$ is the jth decision variable of the best solution found from the previous iteration, ϵ is small value that guides against division by zero, μ is a parameter used to control the search and the MOP coefficient is the Math Optimizer Probability function. This function is given as follows:

$$MOP(t) = 1 - \frac{t^{1/\alpha}}{T^{1/\alpha}} \tag{4}$$

3.1.2 Exploitation Phase

Unlike in the exploration phase of AOA, this phase uses the addition (+) and subtraction (−) operators to move close to the target. From the plot in Fig. 1, the + and − produced highly dense results due to their low dispersion, which means intensifying the search process around a promising solution. This phase of the algorithm highly depends on the result of the MOA function, that is to say, if MOA $> r1$, , (where $r1 \in U(0, 1)$), then exploitation is initiated. The exploitation task with respect to the + and − operators is modelled in Eq. 5

$$X_{i,j}(t+1) = \begin{cases} best(x_j) + (MOP + \epsilon) \times ((ub_j - lb_j) \times \mu + lb_j) & r3 < 0.5 \\ best(x_j) - (MOP + \epsilon) \times ((ub_j - lb_j) \times \mu + lb_j) & otherwise \end{cases} \tag{5}$$

It is worth mentioning that the authors of AOA have carefully designed the μ parameter to maintain randomness as well as strike a balance between exploration and exploitation.

Algorithm 1: Pseudo code of the proposed algorithm.

```
Initialize AOA and DE parameters
Initialize the population of solutions using Eq. 1
Determine best solution (best) from the population
Set t = 1
while (t ≤ T) do
    Calculate MOA and MOP using Eq. 2 and Eq. 4 respectively
    for (i = 1 to N) do
        X_i^new = X_i
        if ( rand ≤ rand ) then
            Mutate dimensions of X_i^new according to Eq. 6
            Apply DE crossover according to Eq. 7
        else
            for (j = 1 to N) do
                Generate values for r1, r2, r3
                if r1 > MOA then
                    if r2 > 0.5 then
                        X_i^new = best_j × (MOP + ε) × ((ub_j − lb_j) × μ + lb_j)
                    else
                        X_i^new = best_j ÷ (MOP + ε) × ((ub_j − lb_j) × μ + lb_j)
                    end if
                else
                    if r3 > 0.5 then
                        X_i^new = best_j + (MOP + ε) × ((ub_j − lb_j) × μ + lb_j)
                    else
                        X_i^new = best_j − (MOP + ε) × ((ub_j − lb_j) × μ + lb_j)
                    end if
                end if
            end for
        end if
        Calculate the fitness of X_i^new
        Accept X_i^new to replace X_i if it has better fitness
        Update best
    end for
    for (k = 1 to K) do
        Set z_{k+1} based on PWLCM of Eq. 9
        Randomly draw 2 solutions (X_m, X_n) from the population, where m ≠ n
        best_new = best_new + (z_{k+1} − 0.5) × (X_m − X_n)
        Accept best_new to replace best if it has better fitness
    end for
    t = t + 1
end while
return best
```

4 The Proposed Algorithm

In this section, we will introduce the DE mutation scheme used to enhance the exploration ability of the AOA, and the chaos strategy used to improve the quality of solutions. We describe these schemes in the subsections below.

4.1 DE Exploration Strategy

Many studies have highlighted premature convergence as the key problem of the AOA. This challenge originates from the algorithm's failure to keep a rich and diverse population during the optimization process. Our proposed method tries to solve this problem by combining AOA with a frequently used DE mutation scheme (DE\rand_to_best\1\bin). This scheme is made up of two terms: the weighted difference between the current and best solution of the population; and the difference between two randomly selected solutions in the population. The full mutation equation is given in Eq. 6. By intuition, the first term takes advantage of the information already known (X_{best}) to induce exploitation, and the second term induces exploration using the distance between two randomly picked solutions. The next step is to refine the mutant vector ($u_i(t)$) using the crossover operation. This is performed for the current solution using a crossover rate to determine the trial vector. The rule states that, for every jth variable of the trial vector and mutant vector that matches, the trail vector keeps the value of the mutant vector, otherwise it retains the value of the corresponding variable of the current solution. The crossover process is mathematically expressed in Eq. 7. In our proposed method we conditioned applying the trial vector by allowing it to update half of the population in every iteration.

$$u_i(t) = X_i(t) + F_i(X_{best}(t) - X_i(t)) + (X_{r1}(t) - X_{r2}(t)) \tag{6}$$

$$w_i(t) = \begin{cases} u_{i,j}(t), & rand_j(0,1) \leq CR || j = j_{rand} \\ X_{i,j}(t), & otherwise \end{cases} \tag{7}$$

$$X_i(t+1) = \begin{cases} w_i(t), & if w_i(t) \leq X_i(t) \\ X_i(t), & otherwise \end{cases} \tag{8}$$

4.2 Chaotic Local Search

Chaos is a form of randomness that is produced by deterministic systems. It is unpredictable and highly sensitive on its choice of the initial value to produce underlying patterns and interconnectedness. The chaos theory has been integrated into many population based metaheuristics in several ways to improve performance. Here, we applied the chaotic local search (CLS) to the best solution produced at the end of each iteration, which serves as a refinement mechanism for the AOA. Among the many available chaotic mapping strategy, we experimentally picked the pair wise linear chaotic map (PWLCM), as it has proved to be computationally effective and efficient. The PWLCM is given in Eq. 9.

$$z_{k+1} = \begin{cases} \frac{z_k}{P} & 0 \leq z_k < p \\ \frac{z_k - P}{0.5 - P} & 0 \leq z_k < 0.5 \\ \frac{1 - P - z_k}{0.5 - P} & 0.5 \leq z_k < 1 - p \\ \frac{1 - z_k}{P} & 1 - P \leq z_k - 1 \end{cases} \tag{9}$$

To ensure a computational efficient process, we desist from using PWLCM with the whole population, but applied it to the best solution found using the following equation:

$$X_{best}^{new} = X_{best,k} + (z_{k+1} - 0.5) \times (X_m - X_n) \tag{10}$$

where X_{best}^{new} is new solution, $X_{best,k}$ is the previous solution at the kth iteration, z_{k+1} is the chaotic value in the range of 0 and 1, and X_m and X_n are solutions randomly picked from the population. The CLS integrated to AOA runs for a number iterations until stopping criteria K is reached, and the population is greedily updated with X_{best}^{new} (Fig. 2).

5 Experiments and Discussion

In this section, we discuss the experimental setting of the study, present and analyse the results obtained.

5.1 *Experimental Settings*

In stochastic optimization, it is important to test any new or modified metaheuristic with different types of functions in order to assess its performance. To accomplish this, we have tested the proposed algorithm (AOADE) alongside other state of the art methods with the CEC 2014 benchmark functions. This benchmark suite contains 30 functions with unimodal, multimodal, hybrid and composite properties. More information about these functions can be found in [29]. For the purpose of achieving a fair comparison, the population size, maximum number of fitness evaluation and number of independent runs for the algorithms tested were set to 30, 10,000 × D, and 50 respectively. Other specific control parameters for these algorithms are

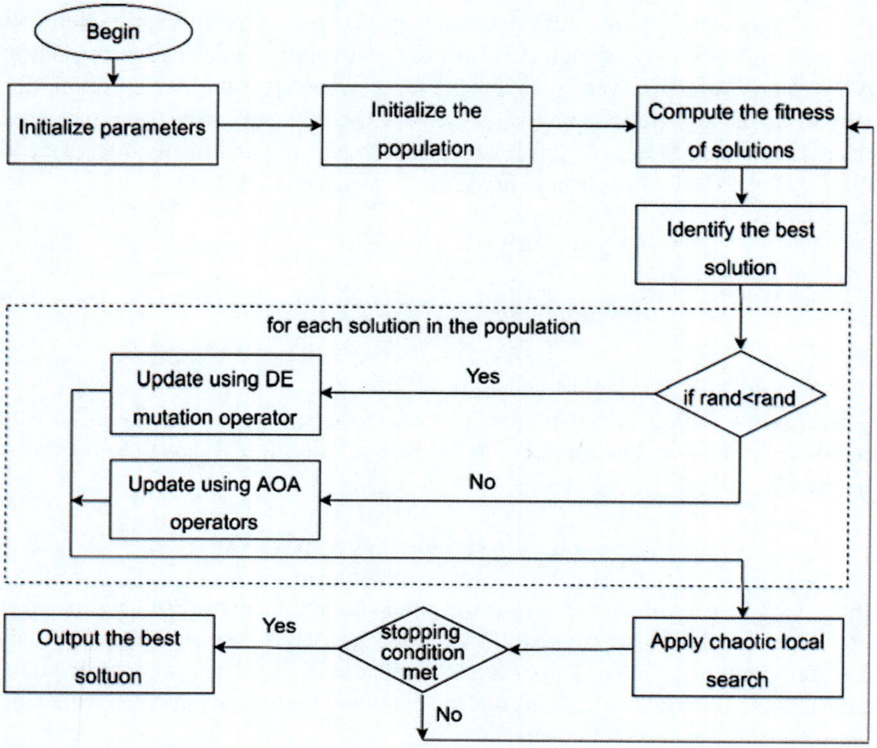

Fig. 2 Flowchart of the proposed method (AOADE)

available in Table 1. The experiments were executed in MATLAB R2020B installed in a windows PC with the following specifications: Intel (R) Core (TM) i5-8250U 2 CPU @ 2.50 GHz, 8 GB RAM.

Table 1 Parameter settings for the AOA, JAYA, IJAYA, DAOA and AOADE

Algorithm	Parameters
AOA	$\alpha = 5$, $\mu = 0.5$
JAYA	No algorithm specific parameter
IJAYA	$\beta = 1.8$
DAOA	$\alpha = 5$, $\mu = 0.5$
AOADE	$\alpha = 5$, $\mu = 0.5$, crossover probability is 0.5, $k = 10$

5.2 Discussion on Solution Accuracy

The result of the experiments conducted in this study for the AOADE and other algorithms (AOA, JAYA, IJAYA, DAOA) are presented in Table 2. The values in Table 2 are organized in a way that each column records the mean and standard deviation of 50 independent runs of one algorithm of the CEC2014 functions. Each value stored in a cell is formatted to keep the average and standard deviation of the independent runs in the left and right side of the ± separator respectively. The bold values represent the best performance on the functions. From Table 2, it can be seen that AOADE statistically outperformed the classical AOA in 27 out of the 30 CEC2014 functions with one tie in F24 and two cases (F23 and F25) where the AOA got better results than AOADE. Going by this, we can easily conclude that our proposed method is better in terms of solution accuracy than the classical AOA. The DAOA which is a recent variant of AOA recorded competitive results in many of the functions tested. In fact, it recorded better result that the AOA in 25 out of 30 functions with no ties. The AOADE and DAOA recorded the same result for one of the composite function (F24) and recorded highly competitive results for at least four functions (F6, F11, F15, and F27). The AOADE also outperformed the DAOA in all the unimodal functions tested which demonstrates the algorithm's advantage in improved exploitation enabled by the chaotic local search mechanism. However the AOADE also showed superior result for most of the multimodal functions (F1, F5, F7, F10, F11, F12, F14 and F15) when compared to DAOAO. This part of the result has confirmed that our proposed method has the ability to jump out of local optimums to avoid the problem of premature convergence which is known with the classical AOA. Also, the JAYA algorithm was no match for our proposed method as AOADE outperforms JAYA in 28 of the 30 CEC2014 benchmark functions.

5.3 Convergence Analysis

Since, the classical AOA is known to converge prematurely on complex functions, we choose to investigate the convergence of our proposed method to ascertain how far this problem has been addressed by AOADE. In Fig. 3, we present the convergence graph for functions of varying categories: unimodal, multimodal and composite functions. From Fig. 3, it can be noticed that for F1, F2 and F12, AOADE tends to steadily converge to the global. This is not the case for the classical AOA which got trapped at an early stage of the optimization.

Table 2 Results of experiments on the CEC2014 functions for 30D

	AOA	JAYA	IJAYA	DAOA	AOADE
F1	8.72e + 08 ± 2.83e + 08	8.47e + 07 ± 2.25e + 07	6.31e + 07 ± 1.87e + 07	5.98e + 06 ± 1.32e + 06	**1.70e + 06 ± 2.36e + 06**
F2	6.59e + 10 ± 6.90e + 09	7.55e + 09 ± 1.18e + 09	4.77e + 09 ± 6.03e + 08	4.75e + 04 ± 3.33e + 03	**3.47e-01 ± 7.50e-01**
F3	7.96e + 04 ± 5.34e + 03	8.10e + 04 ± 1.24e + 04	6.91e + 04 ± 1.07e + 04	1.46e + 05 ± 8.94e + 04	**3.14e + 00 ± 2.31e + 00**
F4	6.89e + 03 ± 2.32e + 03	5.69e + 02 ± 1.29e + 02	4.08e + 02 ± 5.38e + 01	9.42e + 01 ± 3.25e + 01	**6.49e + 01 ± 2.67e + 01**
F5	2.07e + 01 ± 9.78e-02	2.09e + 01 ± 4.71e-02	2.09e + 01 ± 4.97e − 02	2.02e + 01 ± 1.36e-01	**2.00e + 01 ± 1.62e-02**
F6	3.73e + 01 ± 2.52e + 00	3.48e + 01 ± 1.77e + 00	3.39e + 01 ± 1.29e + 00	**1.09e + 01 ± 2.48e + 00**	2.22e + 01 ± 4.61e + 00
F7	5.76e + 02 ± 8.06e + 01	2.59e + 01 ± 5.83e + 00	1.58e + 01 ± 2.80e + 00	1.97e-01 ± 5.37e-02	**8.52e-03 ± 7.61e-03**
F8	3.06e + 02 ± 2.47e + 01	2.29e + 02 ± 1.34e + 01	2.24e + 02 ± 9.93e + 00	**3.22e + 01 ± 1.31e + 01**	1.10e + 02 ± 3.24e + 01
F9	2.48e + 02 ± 2.20e + 01	2.64e + 02 ± 1.85e + 01	2.61e + 02 ± 1.47e + 01	**1.06e + 02 ± 2.56e + 01**	1.63e + 02 ± 2.31e + 01
F10	5.26e + 03 ± 5.16e + 02	5.59e + 03 ± 4.35e + 02	5.68e + 03 ± 3.95e + 02	3.12e + 03 ± 5.21e + 02	**2.25e + 03 ± 6.44e + 02**
F11	5.53e + 03 ± 6.06e + 02	6.91e + 03 ± 3.16e + 02	6.88e + 03 ± 3.12e + 02	3.12e + 03 ± 4.05e + 02	**2.87e + 03 ± 5.83e + 02**
F12	1.17e + 00 ± 2.60e-01	2.44e + 00 ± 2.63e-01	2.49e + 00 ± 2.73e − 01	2.91e-01 ± 9.54e-02	**1.82e-01 ± 9.77e-02**
F13	6.63e + 00 ± 7.67e-01	1.80e + 00 ± 3.59e − 01	1.08e + 00 ± 1.19e − 01	4.92e-01 ± 4.24e-02	**4.01e-01 ± 8.13e-02**
F14	2.10e + 02 ± 3.67e + 01	1.23e + 01 ± 1.82e + 00	4.33e + 00 ± 1.70e + 00	5.53e-01 ± 3.14e-01	**2.87e-01 ± 1.08e-01**
F15	9.40e + 04 ± 3.55e + 04	8.39e + 01 ± 7.03e + 01	5.05e + 01 ± 9.36e + 00	**1.03e + 01 ± 2.24e + 00**	1.10e + 01 ± 2.57e + 00
F16	1.23e + 01 ± 4.16e-01	1.30e + 01 ± 1.70e − 01	1.28e + 01 ± 1.78e − 01	1.29e + 01 ± 2.96e-01	**1.13e + 01 ± 6.43e-01**
F17	1.18e + 07 ± 8.95e + 06	4.69e + 06 ± 1.36e + 06	2.63e + 06 ± 9.76e + 05	2.93e + 05 ± 1.40e + 05	**1.14e + 05 ± 1.04e + 05**
F18	6.41e + 03 ± 3.74e + 03	2.97e + 07 ± 3.19e + 07	1.26e + 07 ± 1.06e + 07	1.67e + 04 ± 7.21e + 03	**1.13e + 03 ± 2.06e + 03**
F19	2.08e + 02 ± 7.45e + 01	3.85e + 01 ± 1.91e + 01	3.78e + 01 ± 3.46e + 01	**1.18e + 01 ± 2.13e + 00**	1.65e + 01 ± 1.23e + 01
F20	9.90e + 04 ± 4.11e + 04	1.16e + 04 ± 3.70e + 03	9.92e + 03 ± 3.69e + 03	9.00e + 04 ± 4.19e + 04	**1.72e + 02 ± 2.20e + 02**
F21	6.81e + 05 ± 4.18e + 05	9.02e + 05 ± 3.08e + 05	6.94e + 05 ± 2.03e + 05	3.79e + 05 ± 2.40e + 05	**3.00e + 04 ± 7.28e + 04**

(continued)

Table 2 (continued)

	AOA	JAYA	IJAYA	DAOA	AOADE
F22	9.67e + 02 ± 2.33e + 02	6.45e + 02 ± 1.38e + 02	5.47e + 02 ± 1.05e + 02	**5.51e + 02 ± 2.07e + 02**	6.86e + 02 ± 2.40e + 02
F23	**2.05e + 02 ± 3.72e + 01**	3.57e + 02 ± 6.69e + 00	3.43e + 02 ± 3.41e + 00	3.16e + 02 ± 4.70e-01	3.07e + 02 ± 2.71e + 01
F24	**2.00e + 02 ± 8.82e-03**	2.61e + 02 ± 4.74e + 00	2.57e + 02 ± 4.04e + 00	2.33e + 02 ± 7.04e + 00	**2.00e + 02 ± 2.75e-03**
F25	**2.00e + 02 ± 0.00e + 00**	2.23e + 02 ± 5.18e + 00	2.16e + 02 ± 2.58e + 00	2.06e + 02 ± 1.31e + 00	2.02e + 02 ± 5.24e + 00
F26	1.76e + 02 ± 3.88e + 01	1.01e + 02 ± 1.70e − 01	1.01e + 02 ± 1.02e − 01	**1.00e + 02 ± 5.57e-02**	1.58e + 02 ± 4.91e + 01
F27	8.84e + 02 ± 5.66e + 02	1.08e + 03 ± 1.96e + 02	9.86e + 02 ± 2.48e + 02	**5.32e + 02 ± 1.08e + 02**	5.97e + 02 ± 2.24e + 02
F28	1.55e + 03 ± 2.05e + 03	1.21e + 03 ± 1.70e + 02	1.13e + 03 ± 6.63e + 01	8.40e + 02 ± 2.47e + 01	**5.20e + 02 ± 5.94e + 01**
F29	3.17e + 08 ± 2.20e + 08	1.57e + 06 ± 3.06e + 06	9.82e + 05 ± 2.07e + 06	5.94e + 06 ± 7.25e + 06	**1.19e + 03 ± 1.19e + 03**
F30	3.31e + 06 ± 3.64e + 06	1.56e + 04 ± 6.41e + 03	1.09e + 04 ± 4.24e + 03	7.19e + 03 ± 3.62e + 03	**1.29e + 03 ± 4.43e + 02**

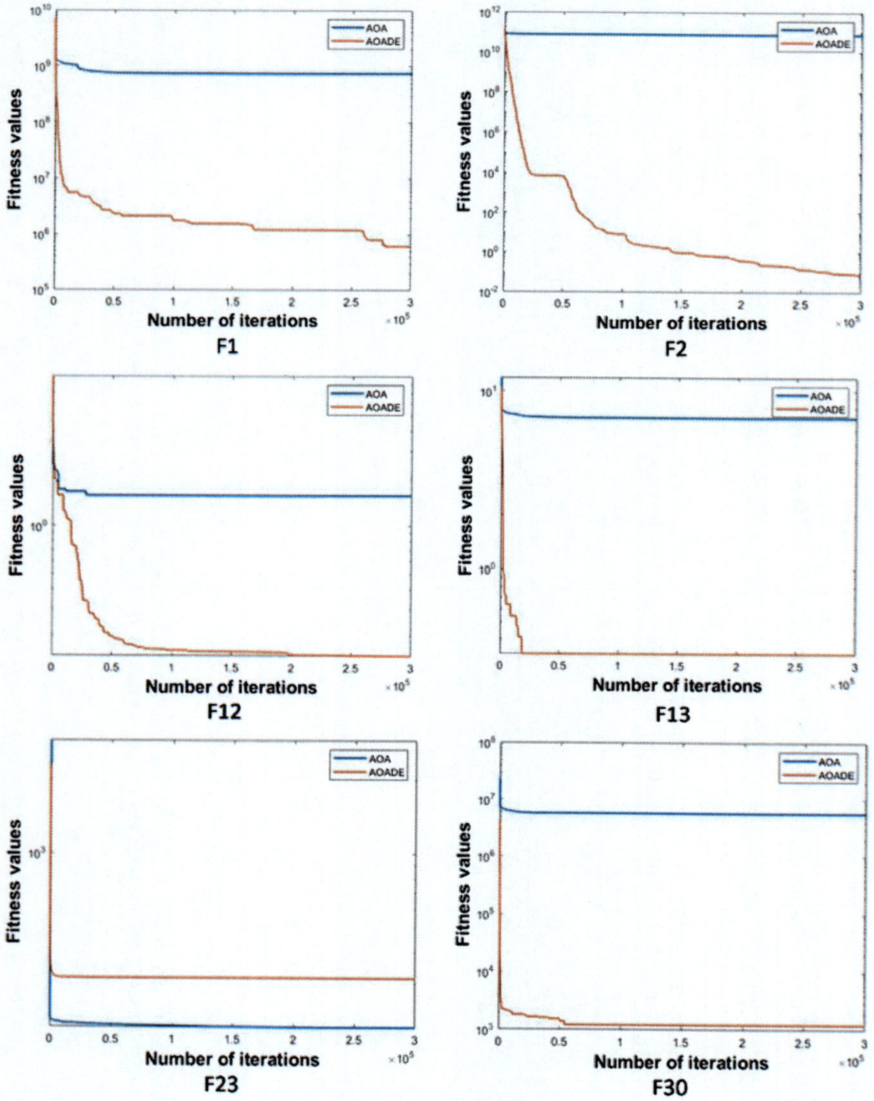

Fig. 3 The convergence graph of AOA and AOADE for selected CEC2014 functions (F1, F2, F12, F13, F23, and F30)

6 Conclusion

In this paper, we proposed a hybrid version of the arithmetic optimization algorithm (AOA) and differential evolution (DE). The new algorithm takes on the local entrapment problem commonly faced by AOA by combining AOA with a modified DE mutation scheme. In addition to this, a chaotic local search mechanism was added to refine the best solution found at the end of each iteration. In order to validate our proposed method, we tested AOADE on the IEEE 2014 benchmark functions and the results obtained were compared against the result of selected state of the art algorithms. The AOADE produced better results than the following algorithms: AOA, JAYA, IJAYA and DAOA in many of the functions tested, and demonstrated the ability to jump out of local optimum. A major limitation of the proposed algorithm is the introduction of new tunable parameters. It is worth mentioning that tuning these parameters to obtain their optimal values for specific applications of the algorithm could be computationally expensive and cumbersome. The future works of the proposed algorithm spans a wide range of application areas. It can be applied to tackle the following problems: selecting subset of features for classification tasks and optimizing the hyper-parameters of neural networks and other machine learning algorithms, just to mention a few.

References

1. G.G. Tejani, V.J. Savsani, V.K. Patel, Adaptive symbiotic organisms search (SOS) algorithm for structural design optimization. J. Comput. Des. Eng. **3**, 226–249 (2016)
2. K.H. Truong, P. Nallagownden, Z. Baharudin, D.N. Vo, A quasi-oppositional-chaotic symbiotic organisms search algorithm for global optimization problems. Appl. Soft Comput. **77**, 567–583 (2019)
3. O.N. Oyelade, A.E. Ezugwu, Characterization of abnormalities in breast cancer images using nature-inspired metaheuristic optimized convolutional neural networks model. Concurr. Comput. Pract. Exp. **34**, e6629 (2022)
4. O.N. Oyelade, A.E. Ezugwu, A bioinspired neural architecture search based convolutional neural network for breast cancer detection using histopathology images. Sci. Rep. **11**, 1–28 (2021)
5. F.E.F. Junior, G.G. Yen, Particle swarm optimization of deep neural networks architectures for image classification. Swarm Evol. Comput. **49**, 62–74 (2019)
6. B. Wang, Y. Sun, B. Xue, M. Zhang, Evolving deep convolutional neural networks by variable-length particle swarm optimization for image classification, in *2018 IEEE Congress on Evolutionary Computation (CEC)* (2018), pp. 1–8
7. C.-L. Huang, J.-F. Dun, A distributed PSO–SVM hybrid system with feature selection and parameter optimization. Appl. Soft Comput. **8**, 1381–1391 (2008)
8. Y.A. Baysal, S. Ketenci, I.H. Altas, T. Kayikcioglu, Multi-objective symbiotic organism search algorithm for optimal feature selection in brain computer interfaces. Expert Syst. Appl. **165**, 113907 (2021)
9. H. Zhu, Y. Jin, Multi-objective evolutionary federated learning. IEEE Trans. Neural Netw. Learn. Syst. **31**, 1310–1322 (2019)
10. A. Al Shorman, H. Faris, I. Aljarah, Unsupervised intelligent system based on one class support vector machine and Grey Wolf optimization for IoT botnet detection. J. Ambient Intell. Hum. Comput. **11**, 2809–2825 (2020)
11. C.E. da Silva Santos, R.C. Sampaio, L. dos Santos Coelho, G.A. Bestard, C.H. Llanos, Multi-objective adaptive differential evolution for SVM/SVR hyperparameters selection. Pattern Recogn. **110**, 107649 (2021)
12. R. Pitakaso, K. Sethanan, N. Srijaroon, Modified differential evolution algorithms for multi-vehicle allocation and route optimization for employee transportation. Eng. Optim. **52**, 1225–1243 (2020)

13. K. Sethanan, R. Pitakaso, Differential evolution algorithms for scheduling raw milk transportation. Comput. Electron. Agric. **121**, 245–259 (2016)
14. M. Abdullahi, M.A. Ngadi, Symbiotic organism search optimization based task scheduling in cloud computing environment. Futur. Gener. Comput. Syst. **56**, 640–650 (2016)
15. D.-H. Tran, J.-S. Chou, D.-L. Luong, Multi-objective symbiotic organisms optimization for making time-cost tradeoffs in repetitive project scheduling problem J. Civ. Eng. Manag. **25**, 322–339 (2019)
16. D.E. Goldberg, J.H. Holland, Genetic algorithms and machine learning. (1988)
17. P. Moscato, On evolution, search, optimization, genetic algorithms and martial arts: towards memetic algorithms, in *Caltech Concurrent Computation Program, C3P Report*, vol. 826 (1989), p. 1989
18. K. Price, R.M. Storn, J.A. Lampinen, *Differential Evolution: A Practical Approach to Global Optimization*: (Springer Science & Business Media, 2006)
19. J. Kennedy, R. Eberhart, Particle swarm optimization, in *Proceedings of ICNN'95—International Conference on Neural Networks* (1995), pp. 1942–1948
20. D. Karaboga, B. Basturk, A powerful and efficient algorithm for numerical function optimization: artificial bee colony (ABC) algorithm. J. Glob. Optim. **39**, 459–471 (2007)
21. X.-S. Yang, S. Deb, Cuckoo search via Lévy flights, in *2009 World Congress on Nature & Biologically Inspired Computing (NaBIC)* (2009), pp. 210–214
22. A. Askarzadeh, A novel metaheuristic method for solving constrained engineering optimization problems: crow search algorithm. Comput. Struct. **169**, 1–12 (2016)
23. A.A. Heidari, S. Mirjalili, H. Faris, I. Aljarah, M. Mafarja, H. Chen, Harris hawks optimization: algorithm and applications. Futur. Gener. Comput. Syst. **97**, 849–872 (2019)
24. A. Faramarzi, M. Heidarinejad, B. Stephens, S. Mirjalili, Equilibrium optimizer: a novel optimization algorithm. Knowl. Based Syst. **191**, 105190 (2020)
25. A. Hatamlou, Black hole: a new heuristic optimization approach for data clustering. Inf. Sci. **222**, 175–184 (2013)
26. S. Mirjalili, SCA: a sine cosine algorithm for solving optimization problems. Knowl. Based Syst. **96**, 120–133 (2016)
27. E. Rashedi, H. Nezamabadi-Pour, S. Saryazdi, GSA: a gravitational search algorithm. Inf. Sci. **179**, 2232–2248 (2009)
28. L. Abualigah, A. Diabat, S. Mirjalili, M. Abd Elaziz, A.H. Gandomi, The arithmetic optimization algorithm, Comput. Methods Appl. Mech. Eng. **376**, 113609 (2021)
29. J.J. Liang, B.Y. Qu, P.N. Suganthan, Problem definitions and evaluation criteria for the CEC 2014 special session and competition on single objective real-parameter numerical optimization, vol. 635 (Computational Intelligence Laboratory, Zhengzhou University, Zhengzhou China and Technical Report, Nanyang Technological University, Singapore, 2013), p. 490
30. X.-D. Li, J.-S. Wang, W.-K. Hao, M. Zhang, M. Wang, Chaotic arithmetic optimization algorithm. Appl. Intell. 1–40 (2022)
31. A.A. Ewees, M.A. Al-qaness, L. Abualigah, D. Oliva, Z.Y. Algamal, A.M. Anter et al., Boosting arithmetic optimization algorithm with genetic algorithm operators for feature selection: case study on cox proportional hazards model. Mathematics **9**, 2321 (2021)
32. S. Mahajan, L. Abualigah, A.K. Pandit, Hybrid arithmetic optimization algorithm with hunger games search for global optimization, Multimed. Tools Appl. 1–24 (2022)
33. L. Abualigah, A.A. Ewees, M.A. Al-qaness, M.A. Elaziz, D. Yousri, R.A. Ibrahim et al., Boosting arithmetic optimization algorithm by sine cosine algorithm and levy flight distribution for solving engineering optimization problems. Neural Comput. Appl. **34**, 8823–8852 (2022)
34. N. Khodadadi, V. Snasel, S. Mirjalili, Dynamic arithmetic optimization algorithm for truss optimization under natural frequency constraints. IEEE Access **10**, 16188–16208 (2022)
35. A. Dahou, M.A. Al-qaness, M. Abd Elaziz, A. Helmi, Human activity recognition in IoHT applications using Arithmetic Optimization Algorithm and deep learning. Measurement **199**, 111445 (2022)

Services Management in the Digital Era—The Cloud Computing Perspective

Aaqif Afzaal Abbasi and Mohammad A. A. Al-qaness

Abstract A digital society consists of smart devices using a number of IT services. These services ensure that quality can be provided to the users for improved user experience. In existing concepts of digital society, smart devices are connected together using a wired or wireless medium of communication. These devices can store data locally or on a cloud which is a shared platform of computational resources. Therefore, the cloud constitutes an important component of the digital society. Cloud systems consist of a network of distributed platforms of storage. The service management of cloud storage and services is often characterized by a legal document known as the service level agreement (SLA). In this paper, a brief discussion is presented to highlight the details of SLAs and their influence on the provisioning of IT services in digital societies.

Keywords Cloud computing

Services Management in the Digital Era—The Cloud Computing Perspective

Aaqif Afzaal Abbasi and Mohammad A. A. Al-qaness

Abstract A digital society consists of smart devices using a number of IT services. These services ensure that quality can be provided to the users for improved user experience. In existing concepts of digital society, smart devices are connected together using a wired or wireless medium of communication. These devices can store data locally or on a cloud which is a shared platform of computational resources. Therefore, the cloud constitutes an important component of the digital society. Cloud systems consist of a network of distributed platforms of storage. The service management of cloud storage and services is often characterized by a legal document known as the service level agreement (SLA). In this paper, a brief discussion is presented to highlight the details of SLAs and their influence on the provisioning of IT services in digital societies.

Keywords Cloud computing · Service level agreement · Smart devices · Storage

1 Introduction

Cloud computing is a cutting edge computing technology, aimed at sharing computational resources. According to the National Institute of Standards and Technology (NIST), cloud computing can be described as one of the most acknowledged technology so far in the field of distributed computing. Its working is dependent on a number of parameters. Cloud computing facilities are omnipresent for users. They provide system access to a common pool of configurable computing assets that can be used to provide improved services execution [1]. The European Community for Programming and Software Services (ECSS) characterizes cloud configuration as the

A. A. Abbasi (✉)
Department of Software Engineering, Foundation University, Islamabad 44000, Pakistan
e-mail: aaqif.afzaal@fui.edu.pk

M. A. A. Al-qaness (✉)
College of Physics and Electronic Information Engineering, Zhejiang Normal University, Jinhua 321004, China
e-mail: alqaness@zjnu.edu.cn

M. Abd Elaziz et al. (eds.), *International Conference on Artificial Intelligence Science and Applications (CAISA)*, Advances in Intelligent Systems and Computing 1441,
https://doi.org/10.1007/978-3-031-28106-8_7

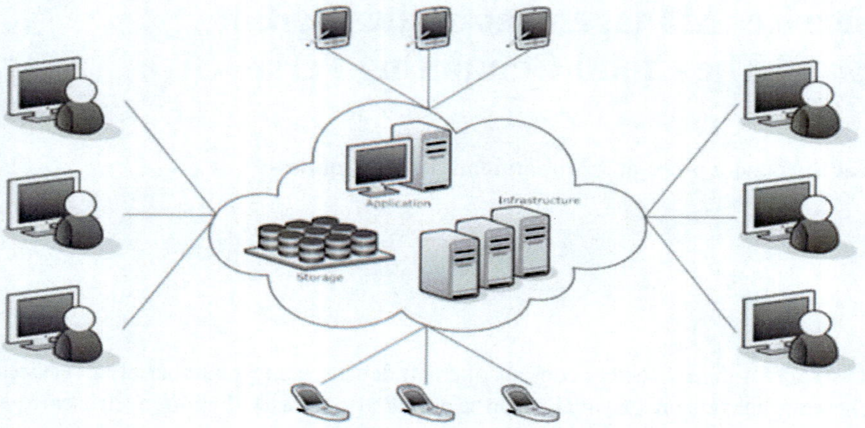

Fig. 1 Infrastructure requirements for cloud deployment

convergence of computational functions by using different virtualization features and functions [1–3]. Therefore, cloud computing is a distributed computing paradigm that provides virtualized services to cloud users to experience computing services from a distributed computing infrastructure. It uses internet infrastructure computational services [4, 5]. Figure 1 illustrates a generic view of cloud infrastructure often used for providing cloud services.

Cloud services are utilized by customers by using online applications. These users consume services by utilizing different heterogeneous equipment such as workstations, mobile devices etc. The changes witnessed in the physical infrastructure of cloud computing include the designing of new cloud applications. However, these services can limit cloud user access on the basis of resource requirements [6]. The biggest advantage to the computing users is that they can use the computing resources whenever required with efforts of purchasing, managing and dealing with the specialized human and communication infrastructure. Due to internet connectivity, users of cloud services can access all resources through the internet. Cloud interfaces are not location dependent however to ensure security features, they use some network and computing-based programs to ensure user location for authenticity [6, 7].

Due to the high trust and reliability features in cloud computing for the utilization of cloud services and features, many cloud orchestration platforms are used. Multi-user-multi-scale modeling is one of the most exciting features of cloud computing. In this paper, a brief description is provided for the cloud computing model based SLAs. It also discusses the services provisioning methods used to manage the user requirements and resource infrastructure.

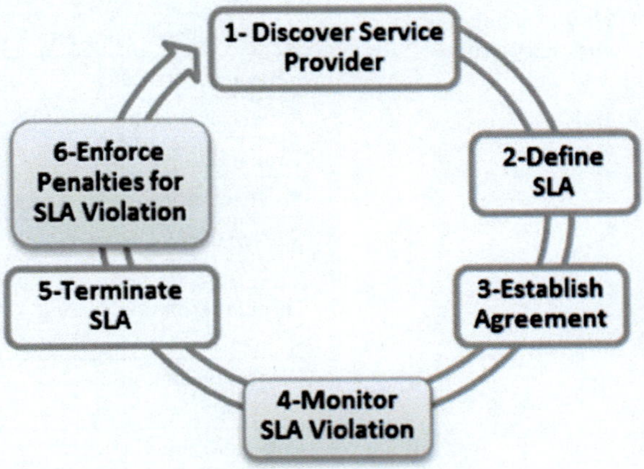

Fig. 2 SLA lifecycle management

2 Related Work

Service provisioning is a very active research domain in services industry. SLA is a legal document that provides the agreed framework between a cloud service provider and a user. It is a formally arranged mutual agreement which depends upon the requirement of the cloud service provider and user. SLAs give a simplistic overview of the cloud users' requirements under predefined scenarios. It also ensures that rules are agreed to meet security issues, legal obligations, certifications and guarantees of the services [6, 7]. Figure 2 illustrates the SLA lifecycle implementation.

SLAs provide an important role in identifying the properties of a cloud network. It provides detailed information about the cloud service provider resource provisioning capabilities by illustrating the list of services. This helps in identifying the parameters through which SLAs can be accommodated [7]. This also helps in developing key security strategies for the cloud environment. SLAs can be distinguished into several categories. This categorization is developed on the basis of customers and their usage. SLA drives an association between the targets of the users and their business-level goals. By mutual agreement among SLA signing parties, the cloud specialist and related personals ensure that the requirements are met on priority. In Table 1, common services provisioning terms are given with a short descriptions.

3 Service Management Challenges Using SLAs

Service management challenges are a big hinderance in open development of cloud infrastructure. In many cases, SLAs are used to incorporate authoritative and specialized features which can guarantee that the cloud processes, features, and functions will be used by particular users. Security threats pertaining to cloud functions would influence numerous users regardless of the number of attacks. These threats are often

Table 1 Common services provisioning terms

Services provisioning terms	Description
Service accessibility	Time of data accessibility
Service availability	Retrieving data from servers
Availability	System uptime
Service change management	Updates for new services
System disaster recovery	Recovery management for the worst-case scenario
Dispute resolution process	SLA violation on either side of the agreement
Process resolution	Newly defined terms if previous agreements are canceled
Data re-location	Datacenter change management
SLA-aware performance	Performance constraints due to strict SLAs
SLA-aware data portability	Data relocation amongst different service providers
Data security	Data encryption and storage

coped by employing infrastructures, applications and data security functions. SLA based privacy preservation not only reduces client data storage issues but also helps in reducing data accessibility challenges by allowing legitimate users to access cloud services.

The data transfer behavior is used to evaluate cloud computing data traffic demands. In this context, SLAs play an important part as they can highlight the traffic movement pattern and resource utilization [5–7]. On the basis of the resource utilization pattern, SLA for effective utilization of cloud services can be adjusted.

The services objectives can be adjusted via SLAs to support various cloud functions. This ensures to evaluate the time during which the network services will be unavailable. Similarly, it can be used to accommodate different applications on the basis of their throughput and response time. In Table 2, various performance metrics are described for services provisioning. Figure 3 illustrates the SLA implementation scenario for a cloud environment.

4 Negotiating on Services in Digital Cities Storage Platforms

Service-level agreements are the base of the cloud industry. Cloud service providers often write these agreements to clearly identify their business requirements. Similar is the case with the cloud service users which agree to these terms for usage of cloud services. It is encouraged for organizations to closely evaluate their company requirements before accepting and incorporating them in a business environment [2,

Table 2 Terms index for services provisioning

Service variables	Description
Rationale	Reason for attaining the objectives for SLAs
Limitations	Procedural formalities for achieving a minimum level of SLA
Validity	The time period for which SLAs are valid
Scope	Service areas covered by SLAs
Actors	Agreeing parties working on SLAs
Objectives	Overall rational for SLA deployment
Penalty	Penalties applied if the minimum level of SLOs are not achieved
Services	Additional services in addition to agreed-upon services
Admin	Processes used to guarantee SLOs and related responsibilities

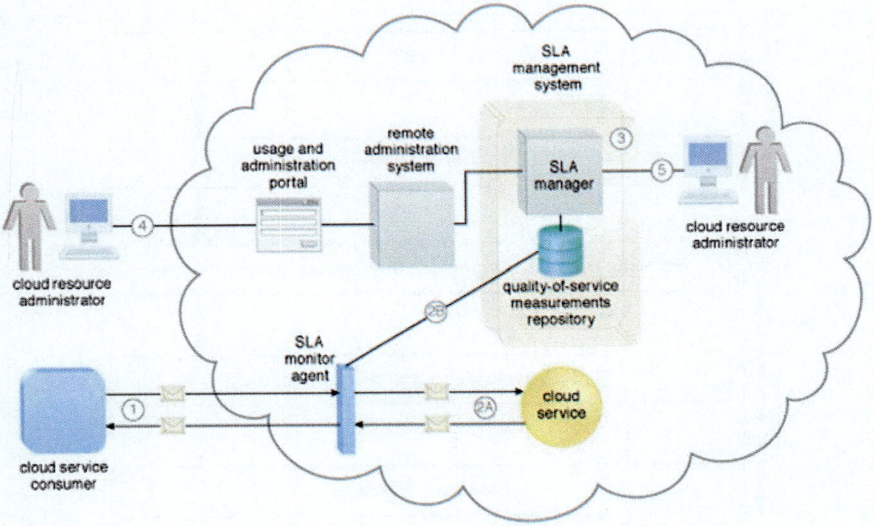

Fig. 3 SLA deployment in a cloud environment

6, 8, 9]. Many computing businesses take advantage of the cloud service providers for managing the data availability function and features of their applications. Negotiating a SLA document in a cloud computing environment might not be an easy way. The reason is the generic nature of services and computation offered by cloud environments. In some aspects, the availability of services in cloud services is considered non-negotiable. The reason being the infrastructure setup and the underlying infrastructure in a cloud computing environment is not very stable. SLA based negotiation

Table 3 SLA performance metrics

Service provider	Discovery mode	SLA definition	Reusable (Yes/No)	Monitoring
Amazon EC2	Manual	Pre-defined	Yes	Allows 3rd party monitoring
Amazon S3	Manual	Predefined	Yes	Allows 3rd party monitoring
MS Azue (compute)	Manual	Pre-defined	Yes	Allows 3rd party monitoring
MS Azue (storage)	Manual	Pre-defined	Yes	Allows 3rd party monitoring

acts as a two-way sword as it also affects the consumers if they violate basic SLA billing agreements on services usage etc. In Table 3, several SLA performance metrics are presented. Figure 4 illustrates a structural view of SLA functional management in a digital society.

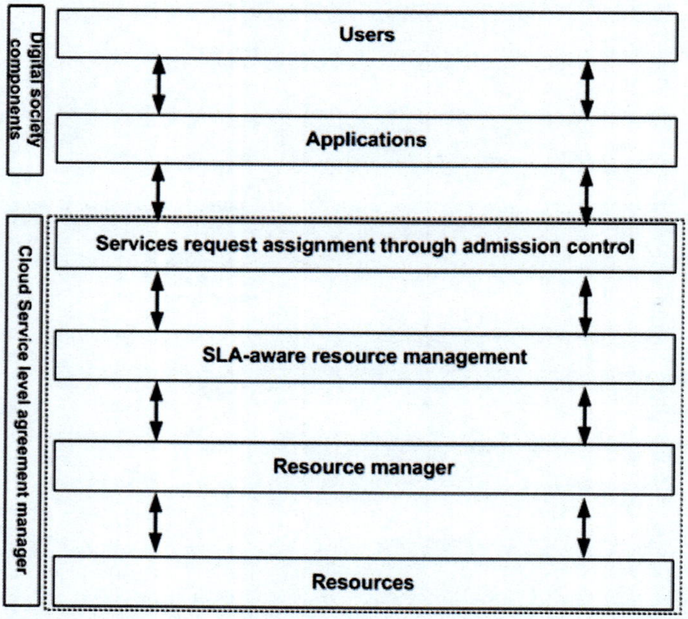

Fig. 4 A structural view of SLA management in a digital society

5 SLA in Changing Times and Requirements

Cloud service providers are often held responsible for SLA development. It is because the SLAs can only be drafted by cloud experts. Its availability, security, and other factors influence bring and charging related information [9–12]. One of the biggest challenges faced in SLA implementation is the inconsistency bought in the cloud infrastructure. The performance issues in SLA compliance is often due to the nature of the infrastructure used by the data centers. The service provider and users of cloud services are both unaware of it and both suffer at the hand of it. Cloud services and applications running on the data center infrastructure constantly share the same equipment multiple times for applications with varying resource needs [13–15]. Applications running on shared time and shifting resource requirements often face SLA challenges. In view of these challenges, the resource allocation issues have been coped-up by using the attest breed of cloud performance monitoring guarantees. These are ensured after a thorough check-up on a day to day basis. These checks ensure performance level guarantees. In Table 4, a literature review is presented on various services provisioning mechanisms.

Table 4 Relevant literature review

Paper ref.	Driver				
	Service management	Services evaluation	Service provisioning	Services guarantees	Services resource management
[16]	✗	✓	✓	✓	✗
[17]	✗	✓	✓	✓	✓
[18]	✓	✓	✓	✓	✓
[19]	✗	✗	✗	✗	✓
[20]	✗	✓	✓	✗	✓
[21]	✓	✗	✓	✓	✗
[22]	✓	✗	✓	✓	✗
[23]	✓	✓	✓	✓	✓
[24]	✗	✓	✗	✗	✓
[25]	✗	✓	✗	✗	✓
[26]	✓	✗	✓	✗	✓
[27]	✓	✗	✓	✓	✓
[28]	✗	✗	✗	✓	✗
[29]	✗	✓	✗	✓	✗
[30]	✓	✓	✓	✗	✓

6 Conclusion

Digital society consists of IT-enabled services which can improve the communication-related constraints of users. Cloud computing has been a major player in the provisioning of resources to digital societies to provide IT storage services. Due to the ever-changing requirements of computing users, cloud computing systems are considered to be the best fit for internet users. The cloud systems utilize the underlying infrastructure and network resources to perform computation tasks. In order to properly allocate these resources, various cloud resource management strategies have been proposed in the literature. SLAs constitute a fundamental component of the computing and service industry. It is a commitment between the service provider and the client for the provisioning of services. In this paper, SLAs were discussed from the digital society's point of view. It also highlighted the terms of services provisioning, negotiation and resource allocation strategies. In future, we plan to associate the areas directly influencing the services provisioning domain.

References

1. A.F.M. Hani, I.V. Paputungan, M.F. Hassan,Renegotiation in service level agreement management for a cloud-based system. ACM Comput. Surv. (CSUR) **47**(3), 51 (2015)
2. K. Lu, R. Yahyapour, P. Wieder, E. Yaqub, M. Abdullah, B. Schloer, C. Kotsokalis, Fault-tolerant Service Level Agreement lifecycle management in clouds using actor system. Futur. Gener. Comput. Syst. **54**, 247–259 (2016)
3. S.A. Baset, Cloud service level agreement.Encyclopedia Cloud Comput. 433 (2016)
4. A.A. Abbasi,A. Abbasi, S. Shamshirband, A.T. Chronopoulos, V. Persico, A. Pescapè, Software-defined cloud computing: a systematic review on latest trends and developments. IEEE Access **7**, 93294–93314 (2019)
5. S. Sharaf, K. Djemame,Enabling service-level agreement renegotiation through extending WS-Agreement specification. Service Orient. Comput. Appl. **9**(2), 177–191 (2015)
6. A.A. Zainelabden, A. Ibrahim, D. Kliazovich, P. Bouvry,On service level agreement assurance in cloud computing data centers, in *IEEE 9th International Conference on Cloud Computing (CLOUD)* (2016), pp. 921–926
7. A.A. Abbasi, M. Hussain,A QoS enhancement framework for ubiquitous network environments. Int. J. Adv. Sci. Technol. **43**, 37–48 (2012)
8. A. Alqahtani, Y. Li, P. Patel, E. Solaiman, R. Ranjan,End-to-end service level agreement specification for iot applications, in *2018 International Conference on High Performance Computing & Simulation (HPCS)* (IEEE, 2018), pp. 926–935
9. E. Rios, E. Iturbe, X. Larrucea, M. Rak, W. Mallouli, J. Dominiak, V. Muntés, P. Matthews, L. Gonzalez,Service Level Agreement-based GDPR Compliance and Security assurance in (multi) Cloud-based systems. IET Softw. (2019)
10. A.A. Abbasi, M.A.A. Al-qaness, M.A. Elaziz, A. Hawbani, A.A. Ewees, S. Javed, S. Kim,Phantom: towards vendor-agnostic resource consolidation in cloud environments. Electronics 8(10), 1183 (2019)
11. A. Stanik, M. Koerner, O. Kao, Service-level agreement aggregation for quality of service-aware federated cloud networking. IET Netw. **4**(5), 264–269 (2015)
12. A.A. Abbasi, M.A.A. Al-qaness, M.A. Elaziz, H.A. Khalil, S. Kim,Bouncer: a resource-aware admission control scheme for cloud services. Electronics **8**(9), 928 (2019)

13. F. Ren, M. Zhang, L. Niu, X. Wang,A concurrent multiple negotiation strategy for service level agreement negotiations in web-based service environments, in *2016 IEEE International Conference on Agents (ICA)* (IEEE, 2016), pp. 1–6
14. Y. Liu, A.A. Abbasi, A. Aghaei, A. Abbasi, A. Mosavi, S. Shamshirband, M.A.A. Al-qaness,A mobile cloud-based eHealth scheme. Comput. Mater. Continua **63**(1), 31–39 (2020)
15. G. Santos-Boada, J.R. de Almeida Amazonas, J. Solé-Pareta,Quality of network economics optimisation using service level agreement modelling. Trans. Emerg. Telecommun. Technol. **27**(5), 731–744 (2016)
16. A.A. Abbasi,S. Shamshirband, M.A.A. Al-qaness, A. Abbasi, N.T. AL-Jallad, A. Mosavi, Resource-aware network topology management framework. Acta Polytechnica Hungrica **17**(4), 89–101 (2020)
17. E. Aubry, T.Silverston, A. Lahmadi, O. Festor, CrowdOut: a mobile crowdsourcing service for road safety in digital cities, in *2014 IEEE International Conference on Pervasive Computing and Communication Workshops (PERCOM WORKSHOPS)* (IEEE, 2014), pp. 86–91
18. F. Duarte, F. de CarvalhoFigueiredo, L. Leite, D.A. Rezende, A conceptual framework for assessing digital cities and the Brazilian index of digital cities: analysis of Curitiba, the first-ranked city. J. Urban Technol. **21**(3), 37–48 (2014)
19. A.A. Abbasi,S. Javed, S. Shamshirband, An intelligent memory caching architecture for data-intensive multimedia applications. Multimed. Tools Appl. 1–19 (2020)
20. J.L. Moutinho, M. Heitor, Digital cities and the opportunities for mobilizing the information society: case studies from Portugal, in *International Digital Cities Workshop* (Springer, Berlin, Heidelberg, 2003), pp. 417–436
21. E. Aubry, T. Silverston, A. Lahmadi, O. Festor, CrowdOut: a mobile crowdsourcing service for road safety in digital cities, in *2014 IEEE International Conference on Pervasive Computing and Communication Workshops (PERCOM WORKSHOPS)* (IEEE, 2014), pp. 86–91
22. A.A. Abbasi, S. Sultana, M.A.A. Al-Qaness, A.Hawbani, S. Javed, S. Kim, Lightweight virtual machine mapping for data centers, in *IEEE International Conferences on Ubiquitous Computing & Communications (IUCC) and Data Science and Computational Intelligence (DSCI) and Smart Computing, Networking and Services (SmartCNS)* (2019), pp. 318–322
23. C. Wang, S. Li, T. Cheng, B. Li, A construction of smart city evaluation system based on cloud computing platform. Evol. Intel. **13**(1), 119–129 (2020)
24. L. Qu, Y. Wang, M.A. Orgun, L. Liu, H. Liu, A. Bouguettaya, CCCloud: context-aware and credible cloud service selection based on subjective assessment and objective assessment. IEEE Trans. Serv. Comput. **8**(3), 369–383 (2015)
25. A.D. de Carvalho Junior, M. Queiroz, G.Essl, Computer music through the cloud: evaluating a cloud service for collaborative computer music applications, in *ICMC* (2015)
26. M. Barcelo, A. Correa, J.Llorca, A.M. Tulino, J.L. Vicario, A. Morell, IoT-cloud service optimization in next generation smart environments. IEEE J. Select. Areas Commun. **34**(12), 4077–4090 (2016)
27. E. Ergazakis, K. Ergazakis, D. Askounis, Y. Charalabidis, Digital cities: towards an integrated decision support methodology. Telematics Inform. **28**(3), 148–162 (2011)
28. I. AlRidhawi, M. Aloqaily, B. Kantarci, Y. Jararweh, H.T. Mouftah, A continuous diversified vehicular cloud service availability framework for smart cities. Comput. Netw. **145**, 207–218 (2018)
29. S.S. Wagle, M. Guzek, P. Bouvry, R. Bisdorff, An evaluation model for selecting cloud services from commercially available cloud providers, in *2015 IEEE 7th International Conference on Cloud Computing Technology and Science (CloudCom)* (2015), pp. 107–114
30. K. Nowicka, Smart city logistics on cloud computing model. Procedia Social Behav. Sci. **151**(Supplement C), 266–281 (2014)

Trip Recommendation Using Location-Based Social Network: A Review

Rizwan Abbas, Irshad Hussain, Gehad Abdullah Amran, Sultan Trahib Alotaibi, Ali A. AL-Bakhrani, Esmail Almosharea, and Mohammed A. A. Al-qaness

Abstract The travel industry is both a significant industry and a well-known recreation movement embraced by millions throughout the planet. One great errand for travelers is to plan and timetable visit schedules that involve the various dazzling Locations of Interest (LOIs) dependent on the extraordinary inclinations of a traveler. The mind-boggling errand of visit agenda proposal is additionally confounded by the need to fuse distinct genuine imperatives like restricted time for visiting, unsure traffic conditions, severe climate, group trips, lining times, and overcrowding. In this learning, we direct thorough writing examination of studies on visit schedule suggestions and present an overall scientific classification for visiting related examination.

R. Abbas (✉) · I. Hussain
College of Software Engineering, Northeastern University, Hunnan, Shenyang 110169, Liaoning, China
e-mail: rizwanabbas@stumail.neu.edu.cn

I. Hussain
e-mail: irshad@stumail.neu.edu.cn

G. A. Amran
Department of Management Science and Engineering, Dalian University of Technology, Ganjingzi, Dalian 116620, Liaoning, China

S. T. Alotaibi (✉)
College of Computing and Information Technology, University of Tabuk, King Faisal Road, Tabuk 71491, Saudi Arabia
e-mail: 421008890@stu.ut.edu.sa

A. A. AL-Bakhrani
Department of Computer Science, Technique Leaders College, 14 October, Sana'a 31220, Yemen

E. Almosharea
Department of Software Engineering, School of Software Technology, Dalian University of Technology, Xuefu, Dalian 116620, Liaoning, China
e-mail: Esmail20@mail.dlut.edu.cn

M. A. A. Al-qaness
College of Physics and Electronic Information Engineering, Zhejiang Normal University, Jinhua 321004, China
e-mail: alqaness@zjnu.edu.cn

M. Abd Elaziz et al. (eds.), *International Conference on Artificial Intelligence Science and Applications (CAISA)*, Advances in Intelligent Systems and Computing 1441,
https://doi.org/10.1007/978-3-031-28106-8_8

We will cover the Location of Interest (LOI) and sequence of LOIs studies that have been done to improve the traveling experiences of users. We talk about the whole cycle of visit schedule suggestion research covering: (i) information assortment and kinds of datasets; (ii) issue plans and suggested calculations/frameworks for individual travelers, gatherings of travelers, and different genuine contemplation; (iii) assessment strategies for looking at a visit to a recommended LOI; (iv) assessment strategies for comparing trip planned route recommendation algorithms; and (v) upcoming bearings and open issues in LOI and trip planned route recommendation research.

Keywords LOI recommendation · LOIs sequence recommendation · Personalized recommendation · Group recommendation · Trip planned route recommendation · Location-based social networks

1 Introduction

The trip recommendation is the Location of Interest (LOI) or Locations of Interest (LOIs) recommendation. The travel industry is a well-known recreation movement attempted by more than 1.18 billion worldwide travelers for every annum [1]. Monetarily, the travel industry is significant, producing more than 284 million positions. The travel industry upholds some 7% of the world's laborers and represents more than US$7.2 trillion in income every year [2]. For all its significance and prevalence, arranging a visit schedule in an unfamiliar city is both testing and tedious because of the need to recognize the charming LOIs and plan visits to these LOIs as an interfacing agenda. Add on to these difficulties is the need to customize the prescribed agenda as per the inclinations of travelers and to plan the schedule dependent on applicable transient and perceptual imperatives, like having a restricted chance to finish the visit and expecting to begin and end close to specific areas (for example the traveler's lodging). Moreover, the traveler needs to discover a schedule that maximizes time spent and the number of LOIs while fulfilling different outing limitations. Location Recommendation: The quick urbanization of urban communities has led to an increase in the number of LOIs to a trip. These LOIs can be restaurants, hotels, beaches, events, cinemas, parks, matches, or even a viewpoint on the road. Location-Based Social Networks (LBSNs) have been a quickly developing field over the most recent years. The volume of data created by LBSNs permits data miners to take out precise user data to offer superior support in end-user applications. The LOIs sequence recommendation is far more challenging than the single LOI recommendation.

Even though travel industry-related data can be obtained from the Internet and travel directs, these assets suggest famously. LOIs or nonexclusive schedules yet, in any case, do not address the particular inclinations of individual sightseers or stick to their different fleeting and perceptual requirements. Additionally, a lot of data accessible builds the test of distinguishing the essential data for the traveler. One mainstream elective is to draw in the administrations of visit offices. However, these travel providers usually only recommend standard package tours that may or may not cater to all traveler's preferences or trip requirements.

To resolve these issues, numerous analysts have considered visit schedule suggestion issues and suggested different calculations for tackling these issues. These issues started from the activities research local areas where the primary center is to plan an ideal way. The proportion of superlatively is commonly founded on a worldwide metric like LOI fame. Hence, there is no personalization dependent on one-of-a-kind user inclinations. With the commonness of cell phones and area-based web-based forums, there has been an expanded accentuation on information-driven ways to deal with visiting schedule proposals to model the inclinations of sightseers better and suggest customized visit schedules that fulfill these inclinations just as other excursion limitations. This study paper centers around such information-driven visit suggestion research, especially on the sorts of information sources utilized, the issue variations formed, the calculations suggested, and the assessment approach utilized.

Firmly identified with the field of visit schedule proposal are the fields of next-area forecast/suggestion [3–7], top-k area proposal [8–13] and travel package/locale proposal [14–17]. Albeit these fields are identified with visit schedule proposals, there are particular contrasts as far as the issue considered. Next-area forecast and proposal plan to recognize the following area that a user will probably visit, dependent on his/her past directions. In contrast, the visit schedule suggestion expects to suggest different LOIs or areas as a direction. Top-k area proposal and travel package/locale suggestion satisfy the basis of suggesting numerous LOIs as a component of a positioned Tour collection. However, an associating schedule has not configured these LOIs by them. Interestingly, the visit agenda proposal has the extra difficulties of preparing a schedule of associated LOIs that appeal to the inclinations of the users while holding fast to the worldly and perceptual imperatives as a restricted time financial plan for visiting and beginning and end at explicit LOIs.

This study centers around works identified with visit agenda proposals and the diverse genuine contemplations fused into this issue.

The rest of this paper is structured as follows. Section 2 explores Surveys and Related studies. Section 3 depicts Traveler's Trip Recommendation. Section 4 Conclusions and Future Work.

2 Surveys and Related Studies

Various parts of the visit suggestions were issued by classification of study and research survey papers. In this segment, we discuss these connected articles and feature the contrasts between this study and the previous articles. The visit suggestion issue is firmly identified with the traveler trip plan issue shrouded in the activities research local area. What is more, subsequently, there have been different study papers [18, 19] focusing on the parts of the issue plan, algorithmic plan, and the intricacy of this issue. Essentially, many visit suggestion issues depend on variations of the Orienteering Problem (OP). What's more [20, 21] give top to bottom conversations on the OP. Specialists, for example, [22] played out a survey of visit proposal frameworks, focusing on applications and frameworks angles like the kinds of interface, the framework functionalities, the suggestion procedures, and the artificial consciousness strategies utilized. Others examined suggestions overall on the spot based on interpersonal organizations [23] and the overall kinds of examination. Flickr photos [24], with a bit of part of their study covering the travel industry-related applications. While these articles provide intriguing conversations into various parts of visit suggestion, our study is different from other research articles: (i) initially, various proposal research surveys as a comprehensive issue. wrapping the entire interaction relative to information assortment, information pre-preparing, visit agenda proposal, experimentation, and assessment; what is more, (ii) second, we give an exhaustive survey of the present status of the quality in visit proposal research.

3 Traveler's Trip Recommendation

This section looks at improvement-based techniques to manage to visit ideas that do not allow for user-customized trips. From that point on, we talk about data-driven visit idea advances toward that join customized subject to user tendencies, the situation in the traffic, and journeying weakness.

3.1 Personalized Trip Recommendations

In the wake of looking at headway-based procedures, We examine data-driven techniques for managing visit propositions that incorporate customization to offer a unique and customized visit plan to each explorer based on their superior tendencies. There are some main investigating issues in such personalization-based methodologies: (i) firmly initiating the inclinations of voyagers; (ii) as part of the suggested visit agenda, combining these trends.

Personalized Location of Interest (LOI) Recommendation Based on Subway Network Highlights and Users' Historical Behaviors- [25] Current recommender systems consistently consider mixed components to recognize and redo the LOI ideas. Genuine lead records and region factors are two kinds of colossal features most profound of ideal circumstances. Regardless, when in doubt, existing methodologies use the Euclidean distance directly, dismissing the traffic factors. What is more, the situation ascribes users' certain practices are not utilized. In this examination, we took the restaurant idea, for example, and proposed a redid LOI recommender system planning the user profile, diner characteristics, users' chronicled lead features, and metro network features. Specifically, the metro network features, for instance, the number of passing stations, holding uptime, and move times, are removed. A dreary neural organization model is used to show user rehearses. Analyses were coordinated

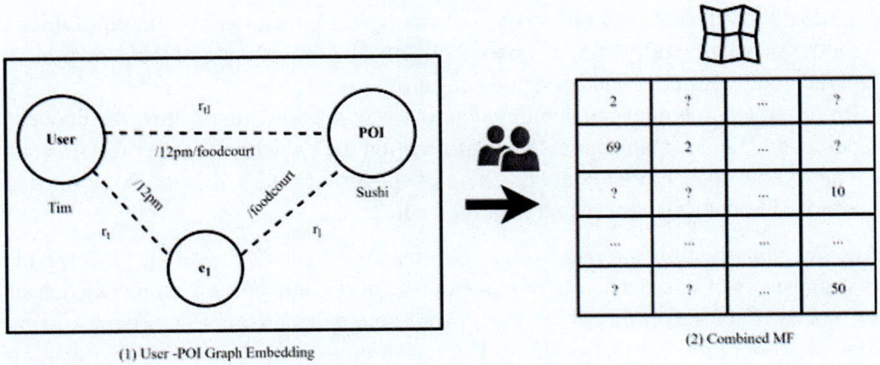

Fig. 1 Overview of GERec model

on an original world dataset, and results show that the recommended system beats the baselines on two estimations. This work recommends the personalized trips to the users [26]. In synopsis, commitments of work incorporate the accompanying:

- Proposed the LOI recommender structure reliant upon solidified features. Took bistro ideas for the assessment case and sorted out ways to solidify user profiles, users' recorded leads, bistro information, and cable car network features.
- Addressed the assortment pattern of user tendencies as time goes on as a fixed-length vector by using a discontinuous neural association.
- Used the Wide and Deep [27] model to expect the scores were given to the bistros by users.
- Direct examinations on a real-world dataset to evaluate the reasonability of our method. Preliminary results demonstrate that our method beats four standard procedures in terms of both Rood Mean Squared Error (RMSE) and Mean Absolute Error (MAE).

Relation Embedding for Personalized Translation-based LOI Recommendation- [28] LOI idea is maybe the main region-based organizations helping people with discovering charming scenes then again benefits. Nevertheless, the ludicrous user LOI system sparsity and the contrasting spatiotemporal setting present challenges for LOI structures, which impact the idea of LOI ideas. To this end, Propose translation-based association embeddings for the LOI idea. Our methodology encodes the transient and geographic information, similarly to semantic substance enough in a low-dimensional association space, using Information Graph Embedding techniques. To relieve the issue of user LOI lattice sparsity, a solidified organization factorization framework depends on a user LOI outline to work on the deduction of dynamic individual tendencies by abusing the side information. Examinations on two certified world datasets show the ampleness of our proposed model. Primary commitments of work incorporate the accompanying:

- To deal with the data sparsity, proposed an original understanding-based LOI idea model to successfully shape a user LOI graph, getting the side information as perceptual, common, and semantic substance.
- Proposed a perceptual-transient association way introducing to show the fleeting, perceptual, and semantic substance information to deal with the idea of LOI ideas.
- They Show their model beats the state of the art LOI idea methodology on two real LBSN datasets, Foursquare and Gowalla.

Visit Recommendation subject to Age, Gender, and Race Cheng et al. [29] intends to suggest visits based on a user's present location and portion nuances such as sex, social class, and caste, which are typically recognized from Flickr photos using a facial recognition estimation [30]. Their visit concept then requires two explicit formulations:

- *Recommending Next LOI* The recommender uses a Bayesian learning model, which also knows the user's present region based on the user section nuances, and Their learning explorer travel model centers on development groupings made up of distinct users with relative fragment credit.
- *Recommending Tour arranged route* They showed visit recommendations concerning the briefest way issue from a starting LOI to target LOI, while moreover joining N diverse LOIs with LOI scores subject to notoriety and plan to the user portion profile. While there is no monetary arrangement based on time or distance (as in normal OPs), The writers embarked on a disciplinary campaign that favored more constrained approaches.

A subsequent work [31] build on [29] by taking into account the size of the social gathering to which a user is traveling, such as individuals, couples, coworkers, and families. They put this theory into practice by employing visual affirmation methodologies to recognize the number of pilgrims in a gathering by counting the number of faces in an image. This work recommends the dynamic trips to a group of users [32].

TripBuilder Algorithm Brilhante et al. [33, 34] suggested the use of the TripBuilder algorithm for organizing changed visit plans for visitors affected by the Generalized Maximum Coverage problem [35]. TripBuilder refers to the process of planning a visit that includes LOIs that boost travelers' unique benefits while adhering to a set visit time and monetary arrangement. There are two phases to TripBuilder:

- *Selection of Sub-directions* The authors employ an evaluation computation as part of the Trip Cover problem to choose a group of sub-headings from LOIs that best suit the explorer's preferences and are within the predetermined time limit.
- *Joining of Sub-directions* The sub-bearings discovered in Step 1 are then combined to create a complete visit plan using a local pursuit calculation as part of the Trajectory Scheduling Problem.

The TripBuilder computation has moreover been made as an online application with a comparative name [36].

TourRecInt Algorithm The TOurRecINT computation [37] intends to propose visit plans with a necessary visit LOI characterization c_m. Hence, this necessary visit LOI order relies upon the LOI arrangement that the voyager is by and large excited about, which the essayist described as the practically sometimes visited LOI class. With the establishment of the essential visit characterization, TOurRecINT relies on an OP variety, which is technically stated as:

$$\sum_{i=1}^{N-1}\sum_{j=2}^{N} x_{i,j}\delta(Cat_i = c_m) \leq 1, \forall c_m \in C \quad (1)$$

where $\delta(Cat_i = c_m) = 1$ if $\text{Cat}_i = c_m$ (LOI I is of class c_m), and 0 in any case. The enhancement work and different limitations are equivalent to the fundamental OP. Visit plans with LOIs and visit ranges tailored to the preferences of individual explorers are recommended [38].

PersTour Algorithm The PersTOur estimation [39, 40] recommends visit plans with LOIs and visit ranges modified to the tendencies of individual explorers. This customization is based on both LOI universality and user tendencies that change over time, This is a general level of user interest in the LOI grouping based on how long an explorer visits the LOI compared to the average time of visit by various visitors. Given that S_u represents explorer u's LOI visit history, voyager u's time-delicate user interest in LOI class c is properly stated as:

$$Int_u^{Time}(c) = \sum_{p_x \in S_u} \frac{\overline{V}_u(px)}{\frac{1}{|T|}\sum_{t \in T} \overline{V}_t(px)} \delta(Cat_{px} = c), \forall c \in C \quad (2)$$

where $\delta(Cat_{\underline{p_x}} = c) = 1$ if $\delta(Cat_{p_x} = c)$, and 0 in any case.

The limit $\overline{V}_t(px)$ shows the typical proportion of time spent by explorer t at LOI p_x, taking into account all the development history of voyager t. From that point on, the PersTOur estimation tries to recommend visit plans like that of the OP, with two standard differentiations, explicitly: (i) PersTOur redesigns for LOI reputation and time-delicate user interest; and (ii) PersTOur surfs a period spending plan subject to both traveling time and a tweaked LOI visit term reliant upon the enjoyment of the user.

Aurigo System Aurigo is a proposed framework that uses an End-to-End mode and a Step-by-Step mode to offer revised timetables [41]. The End-to-End model, like the OP, aims to promote visits with a definite beginning and ending concentrations while increasing LOI acclaim and customer tendencies. Users indisputably give 1–5 star assessments on each LOI order, while LOI acclaim is resolved on Yelp review checks and examinations. In the Step-by-Step manner, the customer chooses an early phase and then iteratively selects the accompanying LOI to visit until he or she is satisfied with the plan that he or she has created.

Photo2Trip System Lu et al. [42] encouraged the Photo2Trip framework, which uses 20 million geo-tagged images and 0.2 million travelogues as the primary inspirations for differentiating notable LOIs, LOI-to-LOI way exposure, and visit ideas. Even more expressly, Photo2Trip achieves these limits by:

- *Identifying standard attractions* MeanShift collecting was used by Photo2Trip to divide images into groups based on their spatial location. The top 10 were then chosen as the best meetings and named based on the travelogues' closest LOI.
- *LOI-to-LOI way disclosure* Because a single user cannot publish all of the images from his or her whole course, Photo2Trip combines diverse parts of the shot to create a single LOI-to-LOI way reliant upon the thickness of the photograph areas and their certified distance.
- *Trip suggestion* Photo2Trip then uses different methods to identify an ideal (standard and captivating) visit that can be completed within a particular time frame, based on the list of LOIs and ways (from Steps 1 and 2).

Setting careful Tour Recommendation Instead of preparing photographs to known LOIs, Majid et al. [43] concludes the space of LOIs and their semantic importance using gathering approaches on geo-tagged photographs. Their system also identifies well-known travel groupings among LOIs and considers the location of the visit idea, such as time, day, and environment. In layout [43] plays out this setting careful visit idea in the going with propels:

- Inferring LOI Locations The P-DBSCAN estimation [44] is first used to organize geo-tagged images into a city's collection of LOI regions. First, the semantic value of LOI is determined using user names and Google Map data.
- Mining persistent travel progressions Next, They used the re-fixSpan computation to arrange geo-tagged images to the identified LOIs to stimulate trip groupings. (for the most part reliant upon [45]) for mining unending travel plans.
- Weather Conditions Determination The scholars then associate each LOI visit with the weather conditions (temperature, dew point, clamminess, squeezing variable, and atmospheric pressure) at the visit using the Wunderground API.
- Recommendation and Letter of Intent User-based aggregate isolating was used to choose LOI interest scores for users. Then the combined probability of LOI interest scores, time, and environment were used to determine the chance of recollecting LOIs for proposed lists.

Visit Recommendation with Time-variety inclinations alternatively OP, Yu et al. [46] recommended a visit idea issue with a starting LOI and visiting time spending plan. Despite this, there was no thought given to a precise LOI target. One main differentiation between this work and others is how Yu et al. proposed time-variety tendencies. For instance, visit get-away objections at the start of the day and eat at a diner around early evening. This work uses the going with propels:

- *Modelling User Interest Preferences* tendencies are shown subject to six-time spans for the term of the day (except for resting time from 12 PM to 08 AM.), and premium levels rely upon visit repeat to LOI classes.
- *Modelling LOI Scores* LOI scores are obtained from a mix of LOI acclaim (based on the number of times that LOI was visited in a given month) and the LOI rating (as consigned by JiePang users to that LOI).

- *LOI Recommendations* The subsequent stage entails identifying LOIs that are appealing to the user and are located near the precise starting point, as well as locating them using user-based aggregate filtering [47].
- *Construction of Tour arranged route* The most recent advancement combines growing a tree from the ground up at a particular LOI and resulting levels based on a list of top-N LOIs [48] for every time. The suggested visit plan is settled ward on a tree crossing. The LOI-to-LOI progress probability relies upon user tendencies, LOI reputation, visiting time, and LOI-to-LOI distances.

Visit Recommendation subject to Time and Seasons In comparison with other visit idea structures, Jiang et al. [49] suggested a method that thinks about tendencies, LOI affirmation costs, LOI opening occasions, Seasons for collection, which they usually get via geo-tagged images and travelog locations. As part of their visit idea structure, they take into account the concept of progress:

- *Extracting LOI Statistics from Travelogs and Photographs* The researchers determine the various characterizations, attestation costs, and opening dates of LOIs based on the titles and depictions of travelogue publications. Timestamps are also utilized to determine how the LOIs were filled during various seasons.
- *Determining User Interest, Cost, Time, and Season Preferences* The authors chose their benefit trends based on the associated names, as well as cost, time, and season trends based on photograph timestamps, using users' posted photographs as development progressions.

MPTR: A Maximal-Marginal-Relevance-Based Customized Trip Suggestion Model [50] With the emergence of location-based services, personalized trip recommendations have gotten much attention recently. Using data from the location-based online community to suggest a single POI or a series of POIs to users is an essential question to answer. The trip recommendation is the process of recommending the latter, which is a difficult study due to the variety of journeys and the complexity of the computations involved. This paper provides a personalized trip recommendation approach based on maximal marginal relevance, which considers both the relevance and diversity of trips while arranging a vacation. An ant colony optimization-based trip planning algorithm is created to plan a journey efficiently. Finally, case studies and experiments show that our strategy is efficient.

The main contributions are following:

- Propose a Maximal-marginal-relevance-based Personalized Tour Recommendation (MPTR) model with the thought of both excursion importance and outing variety in its outing arranging. Extraordinarily, an original estimation technique for LOI closeness is introduced dependent on a predefined class chain of command, and afterward, another assessment procedure is suggested to process trip variety. The consequences are in line with real-life situations.

- Propose an Ant-colony-optimization-based Travel Planning (ATP) the calculation to direct excursion arranging. It can productively plan an excursion, i.e., an arrangement of LOIs.
- To demonstrate the viability of the proposed approach, case studies and tests are conducted.

Up until this point, we have reviewed the visit plan proposition for solitary pioneers and covered improvement-based procedures, personalization-based systems, and web and versatile-based applications. People commonly travel in social groups of varied sizes, such as families and friends, regardless of the situation, and we discuss such endeavors in the surrounding region.

4 Conclusions and Future Work

We have given a sweeping review of the writing in the space of the visit plan idea and highlighted the essential differences between the visit plan proposition and the associated spaces of Operations Research, next-region assumption, top-k region idea, and travel group/nearby idea. We cultivated a logical grouping to portray a general visiting-related assessment. The point-by-point breakdown of the visit plan proposition is subject to various certifiable thoughts like Location of Interest (LOI) reputation, user tendencies, time goals, user economics, transport modes, and traffic conditions. Just as reviewing a big decision of visit plan proposition issues and courses of action, we inspect the different kinds of datasets (geo-tagged electronic discussion, region-based relational associations, and GPS heading follow). Appraisal draws near (reality and heuristic-based estimations, the user consider, and online assessments) that can be used in visit plan idea research.

Considering our survey, we saw a theme of visit recommendations that started from upgrade moves close and towards redid and setting careful methodologies with the transcendence of colossal social data. Notwithstanding how the visit plan proposition has been concentrated late, there is still an exciting assessment orientation to explore. Pushing ahead, we include forthcoming headings that consider diverse new settings and personalization.

Declarations

The authors state that they have no known competing financial interests or personal ties that could appear to have influenced the work disclosed in this study.

References

1. Unwto (2016) United nations world tourism organization (unwto) annual report 2015. Accessed 22 Oct 2017. http://www2.unwto.org/annual-reports
2. World travel and tourism council (2016) 2016 economic impact annual update summary. Accessed 22 Oct 2017. https://www.wttc.org/research/economic-research/economic-impact-analysis/
3. R. Baraglia, C.I. Muntean, F.M. Nardini, F. Silvestri, Learnext: learning to predict tourists movements, in *Proceedings of CIKM'13* (2013), pp. 751–756. https://doi.org/10.1145/2505515.2505656
4. H. Gao, J. Tang, H. Liu, Exploring social-historical ties on location-based social networks, in *Proc.eedings of ICWSM'12*, pp. 114–121. https://www.aaai.org/ocs/index.php/ICWSM12/paper/view/4574
5. D. Lian, V.W. Zheng, X. Xie, Collaborative filtering meets next check-in location prediction, in *Proceedings of WWW'13* (2013), pp. 231–232. https://dl.acm.org/doi/10.1145/2487788.2487907
6. Q. Liu, S. Wu, I. Wang, T. Tan, Predicting the next location: a recurrent model with spatial and temporal contexts, in *Proceedings of AAAI'16* (2016), pp. 194–200. http://www.aaai.org/ocs/index.php/AAAI/AAAI16/paper/view/11900
7. Y. Su, X. Li, W. Tang, J. Xiang, Y. He, Next check-in location prediction via footprints and friendship on location-based social networks. in *Proceedings of MDM'18* (2018), pp. 251–256. https://doi.org/10.1109/MDM.2018.00044
8. K.W.T. Leung, D.L. Lee, W.C. Lee, CLR: a collaborative location recommendation framework based on co-clustering, in *Proceedings of SIGIR'11* (2011), pp. 305–314. https://doi.org/10.1145/2009916.2009960
9. X. Li, G. Cong, X.I. Li, T. Pham, S. Krishnaswamy, Rank-GEOFM: a ranking based geographical factorization method for point of interest recommendation, in *Proceedings of SIGIR'15* (2015), pp. 433–442. https://dl.acm.org/doi/10.1145/2766462.2767722
10. J. Wang, Y. Feng, E. Naghizade, I. Rashidi, K.H. Lim, K.E. Lee, Happiness is a choice: sentiment and activity-aware location recommendation, in *Proceedings of WWW'18* (2018), pp. 1401–1405. https://dl.acm.org/doi/10.1145/3184558.3191583
11. I. Yao, Q.Z. Sheng, Y. Qin, X. Wang, A. Shemshadi, Q. He, Context-aware point-of-interest recommendation using tensor factorization with social regularization, in *Proceedings of SIGIR'15* (2015), pp. 1007–1010. https://dl.acm.org/doi/10.1145/2766462.2767794
12. M. Ye, P. Yin, W.C. Lee, D.I. Lee, Exploiting geographical influence for collaborative point-of-interest recommendation, in *Proc. of SIGIR'11* (2011), pp. 325–334. https://doi.org/10.1145/2009916.2009962
13. Q. Yuan, G. Cong, Z. Ma, A. Sun, N.M. Thalmann, Time-aware point-of-interest recommendation, in *Proceedings of SIGIR'13* (2013), pp. 363–372. https://dl.acm.org/doi/10.1145/2484028.2484030
14. I. Benouaret, D. Lenne, A composite recommendation system for planning tourist visits. in *Proceedings of WI'16* (2016), pp. 626–631. https://ieeexplore.ieee.org/document/7817127/
15. I. Benouaret, D. Lenne, A package recommendation framework for trip planning activities, in *Proceedings of RECSYS'16* (2016), pp. 203–206. https://dl.acm.org/doi/10.1145/2959100.2959183
16. P. Tan, X. Li, G. Cong, A general model for out-of-town region recommendation, in *Proceedings of WWW'17* (2017), pp. 401–410 https://dl.acm.org/doi/10.1145/3038912.3052667
17. M. Toyoshima, M. Hirota, D. Kato, T. Araki, H. Ishikawa, Where is the memorable travel destinations?, in *Proceedings of SOCINFO'18* (2018), pp. 291–298. https://doi.org/10.1007/978-3-030-01159-8_28
18. D. Gavalas, C. Konstantopoulos, K. Mastakas, G. Pantziou, A survey on algorithmic approaches for solving tourist trip design problems. J. Heuristics. **20**(3), 291–328 (2014). https://doi.org/10.1007/s10732-014-9242-5

19. W. Souffriau, P. Vansteenwegen, Tourist trip planning functionalities: state-of-the-art and future, in *Proceedings of ICWE'10* (2010), pp. 474–485. https://doi.org/10.1007/978-3-642-16985-4_46
20. A. Gunawan, H.C. Lau, P. Vansteenwegen, Orienteering problem: a survey of recent variants, solution approaches and applications. Eur. J. Oper. Res. **255**(2), 315–332 (2016). https://doi.org/10.1016/j.ejor.2016.04.059
21. P. Vansteenwegen, W. Souffriau, D.V. Oudheusden, The orienteering problem: a survey. Eur. J. Oper. Res. **209**(1), 1–10 (2011). https://linkinghub.elsevier.com/retrieve/pii/S0377221710002973
22. J. Borrás, A. Moreno, A. Valls, Intelligent tourism recommender systems: a survey. Expert Syst. Appl. **41**(16), 7370–7389 (2014). https://linkinghub.elsevier.com/retrieve/pii/S0957417414003431
23. J. Bao, D.W. Yu Zheng, M. Mokbel, Recommendations in location-based social networks: a survey. Geoinformatica **19**(3), 525–565 (2015). https://www.microsoft.com/en-us/research/publication/recommendations-in-location-based-social-networks-a-survey/
24. E. Spyrou, P. Mylonas, A survey on Flickr multimedia research challenges. Eng. Appl. Artif. Intell. **51**, 71–91 (2016). https://doi.org/10.1016/j.engappai.2016.01.006
25. Y. Danfeng, Z. Xuan, G. Zhengkai, Personalized poi recommendation based on subway network features and users' historical behaviors. Wirel. Commun. Mob. Comput. 3698198:1–3698198:10 (2018). https://www.hindawi.com/journals/wcmc/2018/3698198/
26. R. Abbas, G.M. Hassan, M. Al-Razgan, M. Zhang, G.A. Amran, A.A. Al bakhrani, T. Alfaki, H. Al-Sanabani, S.M.M. Rahman, A serendipity-oriented personalized trip recommendation model. Electronics **11**, 1660 (2022). https://doi.org/10.3390/electronics11101660
27. H.T. Cheng, l. Koc, J. Harmsen, Wide & deep learning for recommender systems (2016). https://arxiv.org/abs/1606.07792
28. W. Xianjing, F.D. Salim , R. Yongli , P. Koniusz, Relation embedding for personalised translation-based poi recommendation. PAKDD (1), 53–64 (2020) https://www.researchgate.net/publication/341243343_Relation_Embedding_for_Personalised_Translation-Based_POI_Recommendation
29. A.J. Cheng, Y.Y. Chen, Y.T. Huang, W.H. Hsu, H.Y.M. Liao, Personalized travel recommendation by mining people attributes from community-contributed photos, in *Proceedings of MM'11* (2011), pp. 83–92 https://dl.acm.org/doi/10.1145/2072298.2072311
30. P. Viola,M. Jones, Rapid object detection using a boosted cascade of simple features, in *Proceedings of CVPR'01*, pp. 511–518 (2001). https://doi.org/10.1109/CVPR.2001.990517
31. Y.Y. Chen, A.J. Cheng, W.H. Hsu, Travel recommendation by mining people attributes and travel group types from community-contributed photos. IEEE Trans. Multimed. **15**(6), 1283–1295 (2013). https://doi.org/10.1109/TMM.2013.2265077
32. R. Abbas, G.A. Amran, A. Alsanad, S. Ma, F.A. Almisned, J. Huang, A.A. Al-Bakhrani, A.B. Ahmed, A.I. Alzahrani, Recommending reforming trip to a group of users. Electronics **11**, 1037 (2022). https://doi.org/10.3390/electronics11071037
33. I. Brilhante, J.A. Macedo, F.M. Nardini, R. Perego, C. Renso, Where shall we go today? planning touristic tours with tripbuilder, in *Proceedings of CIKM'13* (2013), pp. 757–762. https://dl.acm.org/doi/10.1145/2505515.2505643
34. I.R. Brilhante, J.A. Macedo, F.M. Nardini, R. Perego, C. Renso, On planning sightseeing tours with tripbuilder. Inf. Process. Manag. **51**(2), 1–15 (2015). https://doi.org/10.1016/j.ipm.2014.10.003
35. R. Cohen, I. Katzir, The generalized maximum coverage problem. Inf. Process. Lett. **108**(1), 15–22 (2008). https://linkinghub.elsevier.com/retrieve/pii/S0020019008000896
36. I. Brilhante, J.A. Macedo, F.M. Nardini, R. Perego, C. Renso, Tripbuilder: a tool for recommending sightseeing tours, in *Proceedings of ECIR'14* (2014), pp. 771–774. https://www.researchgate.net/publication/280253984_TripBuilder_A_Tool_for_Recommending_Sightseeing_Tours
37. K.H. Lim, Recommending tours and places-of-interest based on user interests from geo-tagged photos, in *Proceedings of SIGMOD'15 Ph.D. Symposium* (2015), pp. 33–38. https://dl.acm.org/doi/10.1145/2744680.2744693

38. K. Taylor, K.H. Lim, J. Chan, Travel itinerary recommendations with must-see points-of-interest, in *Proceedings of WWW'18* (2018), pp. 1198–1205. https://dl.acm.org/doi/10.1145/3184558.3191558
39. K.H. Lim, J. Chan, C. Leckie, S. Karunasekera, Personalized tour recommendation based on user interests and points of interest visit durations, in *Proceedings of IJCAI'15* (2015), pp. 1778–1784. http://ijcai.org/Abstract/15/253
40. K.H. Lim, J. Chan, C. Leckie, S. Karunasekera, Personalized trip recommendation for tourists based on user interests, points of interest visit durations and visit recency. Knowl. Inf. Syst. **54**(2), 375–406 (2018). https://doi.org/10.1007/s10115-017-1056-y
41. A. Yahi, A. Chassang, I. Raynaud, H. Duthil, D.H.P. Chau, Aurigo: an interactive tour planner for personalized itineraries, in *Proceedings of IUI'15* (2015), pp. 275–285. https://dl.acm.org/doi/10.1145/2678025.2701366
42. X. Lu, C. Wang, J.M. Yang, Y. Pang, I. Zhang, Photo2trip: generating travel routes from geo-tagged photos for trip planning, in *Proceedings of MM'10* (2010), pp. 143–152. https://doi.org/10.1145/1873951.1873972
43. A. Majid, I. Chen, H.T. Mirza, I. Hussain, G. Chen, A system for mining interesting tourist locations and travel sequences from public geo-tagged photos. Data Knowl. Eng. **95**, 66–86 (2015). https://doi.org/10.1016/j.datak.2014.11.001
44. S. Kisilevich, F. Mansmann, D. Keim, P-DBSCAN: a density based clustering algorithm for exploration and analysis of attractive areas using collections of geo-tagged photos, in *Proceedings of COM.GEO'10* (2010), p. 38. https://doi.org/10.1145/1823854.1823897
45. J. Han, J. Pei, B. Mortazavi-Asl, H. Pinto, Q. Chen, U. Dayal, M.C. Hsu, Prefixspan: mining sequential patterns efficiently by prefix-projected pattern growth, in *Proceedings of ICDE'01* (2001), pp. 215–224. https://ieeexplore.ieee.org/abstract/document/914830
46. Z. Yu, H. Xu, Z. Yang, B. Guo, Personalized travel package with multi-point-of-interest recommendation based on crowdsourced user footprints. IEEE Trans. Hum. Mach. Syst. **46**(1), 151–158 (2016). https://doi.org/10.1109/THMS.2015.2446953
47. V.W. Zheng, Y. Zheng, X. Xie, Q. Yang, Collaborative location and activity recommendations with GPS history data, in *Proceedings of WWW'10* (2010), pp. 1029–1038. https://dl.acm.org/doi/10.1145/1772690.1772795
48. Y. Zheng, I. Zhang, X. Xie, W.Y. Ma, Mining interesting locations and travel sequences from GPS trajectories, in *Proceedings of WWW'09* (2019), pp. 791–800. https://dl.acm.org/doi/10.1145/1526709.1526816
49. S. Jiang, X. Qian, T. Mei, Y. Fu, Personalized travel sequence recommendation on multi-source big social media. IEEE Tran. Big Data **2**(1), 43–56. https://doi.org/10.1109/TBDATA.2016.2541160
50. W. Luan, G. Liu, C. Jiang, M. Zhou, MPTR: A maximal-marginal-relevance-based personalized trip recommendation method. IEEE Trans. Intell. Transp. Syst. **19**(11), 3461–3474 (2018). https://ieeexplore.ieee.org/document/8306447

A Multimodal Spam Filtering System for Multimedia Messaging Service

Insaf Kraidia, Afifa Ghenai, and Nadia Zeghib

Abstract Filtering systems that detect spam by a single modality have increased over the past few years for Short Message Services (SMS). With the growth of marketing via Multimedia Messaging Services (MMS), a spammer can evade detection by injecting multimodal junk information into MMS to decrease systems recognition based on a single modality. Due to this situation, we propose in this paper, a powerful spam MMS filtering system, with a multi-modality fusion technique, it can detect spam whether it is hidden in text or images. The architecture of the proposed system combines along Short-Term Memory (LSTM) model and a Convolutional Neural Network (CNN) model to filter spam MMS. Based on the text and image portions of an MMS, the LSTM and the CNN calculate two classification probabilities. Afterwards, to determine whether the MMS is spam or not, these values are incorporated into a fusion model. Based on the results of the experiments, the overall accuracy is 98.56% for the MMS dataset, 96.97% for the text-based dataset, and 97.98% for the image-based dataset. We have compared the proposed system with some relevant ones and we have found that it performs better in many criteria such as precision, f1-score, recall, and accuracy and gives an improvement between 1 and 5% in smishing detection.

Keywords Multimedia messaging service · Short messaging service · Multi-modality · Spam filtering system · CNN · LSTM

I. Kraidia (✉) · A. Ghenai · N. Zeghib
LIRE Laboratory, Constantine2 – Abdelhamid Mehri University, Constantine, Algeria
e-mail: insaf.kraidia@univ-constantine2.dz

A. Ghenai
e-mail: afifa.ghenai@univ-constantine2.dz

N. Zeghib
e-mail: nadia.zeghib@univ-constantine2.dz

M. Abd Elaziz et al. (eds.), *International Conference on Artificial Intelligence Science and Applications (CAISA)*, Advances in Intelligent Systems and Computing 1441,
https://doi.org/10.1007/978-3-031-28106-8_9

1 Introduction

Marketing through messaging services is a continuous process of communicating business, promotions, sales, news, or more relevant data to customers via mobile messages. In general, the spammer message the user; this message contains the email identifiers or phone numbers of the spammer or contains URLs to spam user interfaces that ask the user to enter their credentials. Users expose their crucial data (such as passwords, credit card information, etc.) using user interfaces.

Several Deep Learning (DL) and Machine Learning (ML) spam filtering systems are suggested to classify and filtering SMS. Over time, MMS marketing is a steadily growing and thriving industry. In this area, the stats confirmed that more than 300 billion worldwide MMS messages are sent each year (More than 90% of messages are opened in 3 min), and according to some brands, MMS shows 300% more engagement than SMS-only messages [1]. That's why the spammer proposed new scenarios for deceiving the existing SMS filtering. An MMS attack contains multimedia (video and image) embedded with text, By using this technique, spammers can bypass text-based filters. This multimedia tricks users into clicking on it, leading to unsafe websites, causing malware infections, or containing some spam text pieces of information. Figure 1 shows various examples of attack scenarios via SMS and MMS. (a) represents a simple spam SMS, (b) contains a spam MMS with malicious text and a legitimate image, whereas, (c) contains a spam MMS with legitimate text and malicious images.

People may think that spam MMS and SMS filtering systems are similar, but by the presence of multimedia content in the MMS, the SMS approach needs to change to be more effective with the new modality of data. Thus, to effectively identify MMS spam, it is extremely important to analyze the multimedia behavior (specific images) present in the MMS. To the best knowledge, we first highlighted this system in spam MMS filtering domains.

In particular, Convolutional Neural Network (CNN) is more accurate for image classification tasks compared to other methods. On the other hand, Natural language processing commonly uses LSTM due to its time and memory properties [2]. Therefore, CNNs and LSTMs are adapted to deal with the data relating to images and text of the same MMS, respectively. Preserving the research line, this study presents a

Fig. 1 Examples of attack scenarios via SMS and MMS

multimodal architecture that uses LSTM and CNN to generate feature vectors from both images and text of the same MMS. The generated vectors are combined into an assembled model using the logistic regression layer that classified an MMS as spam or ham. Following are the remaining sections of this paper: The related work is summarized in Sect. 2. A description of the system can be found in Sect. 3. Findings from the experiment are discussed in Sect. 4. Section 5 concludes and discusses some potential lines for future research.

2 Related Work

For a long time, spam detection approaches have been extensively studied. Existing approaches are categorized into two broad areas: multi-modal and uni-modal. The first category elaborates on spam classification including both text and image [2]. The second category consists of image-based and text-based categories. For the text-based approach, the authors in [3] proposed a model named "Smishing Detector" to detect SMS spam, using the Naive Bayes (NB) algorithm. In the first module of this model, malicious parts of text messages are analyzed and determined. The next module is for looking at the URL in the SMS. The last module focuses on determining if a spam file is uploaded from the URL. The final module is devoted to the analysis of the site's source code related to messages. The performance evaluation of this approach shows a rate of 96.29%. The authors in [4] suggested a model called SmiDCA to detect smishing messages by ML technique. By the use of a correlation algorithm, the suggested model extracts 39 features from spam SMS. The experimental results achieved an accuracy of 96.40% with the Random Forest algorithm (RF). In [5], four data sets were added to the universal model by fine-tuning the pre-trained BERT-uncased for spam detection. The obtained hyper-parameters from individual models helped train the final model by combining all the samples. The performance evaluation of this approach shows a rate of 98%.

For the image-based approach, the literature presents many works exploiting image features as training data. The authors in [5] used Support Vector Machine (SVM) and Principal Component Analysis (PCA) to detect malicious images where the feature set consists of Twenty-one image properties. Each feature is assigned a weight under its contribution to a linear SVM classification. Based on these weights, the authors accomplish diverse experiments mainly affecting feature selection and feature reduction. These experiments are executed on the Image Spam Hunter dataset [6]. The authors of [7] recommended a CNN for image spam classification. The proposed network which has three convolutional layers and a dropout layer to avoid overfitting was trained over the ISH dataset. The authors in [2] combined CNN and LSTM models on Enron, Spam Archive, and Personal Image datasets. The result for image spam detection was 92.1% according to the accuracy metric.

We note that the proposed works in the spam message service domain only studied text or images, so it makes sense to propose a new filtering system to overcome the multimodality problem in the classification of MMS.

3 The Proposed System

The following sections provide an overview of our system which is called MMS-FS (Multimedia Messaging Service Filtering System) which is a spam detector that can filter MMS. But first, we present the principal sub-models of our system which are: the image classification model to filter MMS images, the text classification model to classify MMS text, and the text-image fusion model to fusion the final predictions.

3.1 Image Classification Model

The CNN model consists of 3 convolution layers, where the first convolutional layer consists of 32 filters, the second one contains 64 filters, and the third one contains 128 filters. After the first and second convolutional layers, a max-pooling layer (size 2) is applied. Following the last convolution layer, a dense layer containing 138 neurons is used after flattening the output of the dropout regularization with the ReLU activation function. Lastly, a sigmoid activation function is used with a dense layer of two neurons. A brief overview of the CNN model architecture, the specification of each layer, and the output shapes are shown in Fig. 2.

CNN hyperparameter values are optimally selected using a grid search optimization technique, such as the SGD optimization algorithm, batch sizes equal to 20, 100 epochs, and a learning rate of 0.01. Detailed descriptions of the CNN algorithm are provided in [8]. Figure 3 illustrates the whole MMS image classification process. Using the image "img" as input, the probability value of classification "r" for the MMS image is obtained.

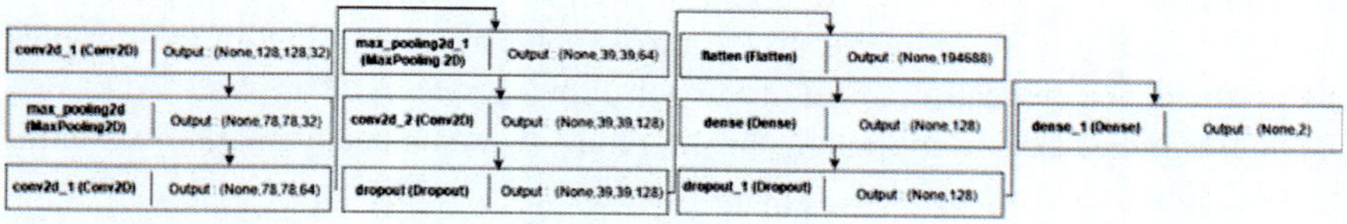

Fig. 2 CNN model architecture

Algorithm 2: Image Spam Classification Algorithm

Input: Image *img*, size 128 × 128 RGB

Output: Image spam classification probability value *r*

Input *img* to the 3 convolutional layers, getting the results *g*, $g=(g_1,\cdots,g_{128})$;

Input g to the first 2 FC layers, which contain 64 and 32 neurons, gaining feature vector *c*, $c = (c_1,c_2,\cdots,c_{128})$;

Input c to the last FC layer, which contains two neurons, using a sigmoid activation function to obtain classification probability value *r*;

return *r*;

Fig. 3 Classification algorithm for malicious images

3.2 Text Classification Model

The general theory of a Recurrent Neural Network (RNN) demonstrates some promising properties but some weaknesses in the remembering of long-term memories. To solve these problems, LSTM is used as an alternative to RNN and aims to add cell states for memorized or forgotten data. The cell states include structures named cell gates which consist of four gates: Input gate, memory-cell state gate, output gate, and forget gate, as shown in Fig. 4.

Our LSTM model consists of two LSTM layers, a word-embedded layer and a dense layer. Following are the steps for calculating the class likelihood value according to the text portion of an MMS:

- First, we use the Word2Vec (W2V) method to obtain the word embedding (WE) representation.
- After that, features are automatically extracted from the text data via the developed two LSTM layers.
- Finally, by applying the softmax activation function to a dense layer, we can obtain a classification probability value.

LSTM hyperparameter values are optimally selected using a grid search optimization technique, such as the Adam optimization algorithm, batch sizes equal to 32, 100 epochs, and a 0.2 learning rate.

The LSTM unit receives the sentences simultaneously with the previous unit's output. All essential features are kept in this unit and this process is repeated with each new input sentence. From the dense and LSTM layers, the model can get the classification probability value "r" of the text portion. The LSTM algorithm details can be found in [9] (Fig. 5).

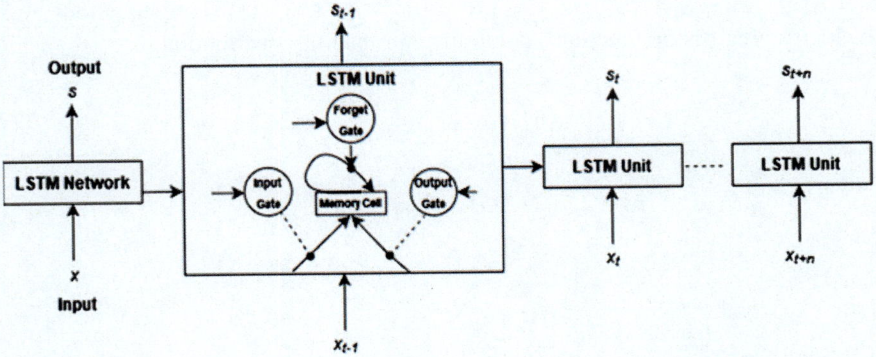

Fig. 4 LSTMs architecture

Algorithm1: Text Spam Classification Algorithm
Input: Text Document *txt*
Output: Text spam classification probability value *r*
Input *txt* into the word2vec toolkit to get the word vector *x*, $x = (x_1,..,x_n)$;
Input *x* to the first LSTM layer, get the text feature vector $h = (h_1,..,h_{64})$;
Input *h* to the second LSTM layer and get more abstract text feature vector *k*, $k = (k_1, k_2,..,k_{32})$;
Input *k* to FC layer and use SoftMax activation function to gain the text classification probability value *r*;
return *r*;

Fig. 5 Text spam classification algorithm

3.3 Text-Image Fusion Model

In this subsection, to obtain the highest accurate classification probability value of the MMS. we combine the classification probability value of an MMS text part with the classification probability value of the same MMS image part. For That, a fusion model is proposed following these steps:

- To get the first version of the feature vector g, $g \in R^{1\times4}$. We have to merge two class likelihood values of the LSTM and CNN models.
- After the concatenation of both modalities, a fully connected (FC) layer with sixty-four (64) neurons is used to obtain an extensive feature vector.
- An extensive feature vector is applied to the logistic layer, consisting of two neurons, to determine whether the MMS is spam. As an activation function, logistic regression is used.
- Adam optimizer is used with a mini-batch equal to 32 for training the model. To avoid overfitting, early-stopping conditions are used, and as an activation function, the ReLU function is chosen.

The classification probability values inputting to the *Text-Image Fusion model* is $K = \{(s_1, y_1), (s_2, y_2), \ldots, (s_v, y_v)\}$, $s_i \in R^{1\times4}$, $y_i \in \{0, 1\}$. These equations give the logistic regression function's conditional probability distribution:

$$P(Y = 1|s) = \pi(s) = \frac{e^{-w^T \cdot s}}{1 + e^{-w^T \cdot s'}} \tag{1}$$

$$P(Y = 0|s) = 1 - \pi(s) = \frac{1}{1 + e^{-w^T \cdot s'}} \tag{2}$$

3.4 The Proposed System

An overview of the flowchart of our system is illustrated in Fig. 6. MMS-FS contains five specific steps to identify spam MMS:

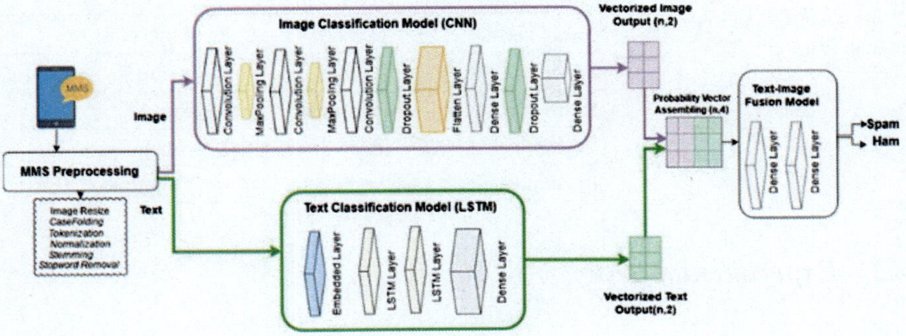

Fig. 6 MMS-FS architecture

- Image preprocessing: Changes are made to the size of the image 128*128 (pixels). This is done because of the input image size variety.
- Text preprocessing: Contains the multiple following steps:
  - Case Folding: Consists of converting uppercase to lowercase letters.
  - Tokenization: Consists of decomposing sentences into one or more words and eliminating delimiters such as dots (.), commas (,), and spaces.
  - Normalization: Consists of converting a token into its base format. In this process, the inflectional form of a word is removed.
  - Stemming: Is needed to group words that have similar meanings but have different forms because they have different affixes.
  - Stop word Removal: Consists of removing words or features that are not important and often appear in text documents such as conjunctions.
- Get the optimal classifiers: To train and optimize the text and image Classification Models, text-based and image-based datasets respectively are used.
- To calculate the probability of classifying MMS as spam, the image and text datasets are re-entered into the optimal CNN and LSTM models.
- Get the final prediction: The Text-Image Fusion model can average the prediction results of all sub-modules.

4 Experiments Results and Discussion

We present here the results of a variety of experiments. Then, we discuss the datasets that we used to assess our system and present the measures to evaluate the prediction rate.

Table 1 MMS dataset probabilities

Image class	Text class	Final class
Spam	Spam	Spam
Spam	Legitimate	Spam
Legitimate	Spam	Spam
Legitimate	Legitimate	Legitimate

4.1 Experimental Setup

We implement our proposed system on Google Cloud GPU for image classification and Google Cloud CPU for text classification using Google Colaboratory. We used Python 3 and Deep Learning libraries, such as Keras 2.1.6 and TensorFlow 1.7.0.

4.2 Datasets

In our experiments, two publicly available datasets are used and one dataset is built.

- **Image Spam Hunter (Image-based dataset)** [6]: It is a public dataset that contains malicious and legitimate images in JPEG format. ISH includes 810 legitimate images that are randomly collected and 926 malicious images from real spam content.
- **SMS Spam Collection (Text-based dataset)** [10]: This dataset includes 5574 text messages, of which 747 SMS messages are malicious and 4827 are legitimate.
- **MMS Dataset**: As there is no public MMS dataset that consists of both text and image data, it is necessary to create one, The previous datasets were combined as follows: 1202 text messages and 4808 images are formed into 4808 MMS messages (the combination is not randomly). Table 1 shows the building probabilities that we used based on [7] and [11].

4.3 Evaluation Criteria and Validation Method

Regarding quality measures, the most used measures that are chosen in the literature are precision, recall, F1 score, and accuracy. These measures are defined as follows:

$$Accuracy = \frac{\text{True Negative} + \text{True Positive}}{\text{True Positive} + False\ Negative + \text{True Negative} + \text{False Positive}} \tag{3}$$

$$\text{Precision} = \frac{True\ Positive}{False\ Positive + True\ Positive} \tag{4}$$

Table 2 The confusion matrix of prediction results using multimodal data (MMS dataset)

	Prediction		Number of each category
	Normal	Spam	
Normal	163 (TN)	10 (FP)	173
Spam	7 (FN)	935 (TP)	942
Testing data			1115

$$Recall = \frac{True\ Positive}{False\ Negative + True\ Positive} \quad (5)$$

$$F1 - score = \frac{2 * Recall * Precision}{Recall * Precision} \quad (6)$$

The definitions of specific meanings of metrics are defined as follows:

- True Negative: Present the number of legitimate MMS correctly classified.
- True Positive: present the number of correctly classified spam MMS.
- False Negative: Present the number of misclassified spam MMS.
- False Positive: Present the number of legitimate MMS misclassified.

4.4 Experimental Results

A summary of our evaluation results is presented in this section, along with an analysis and discussion of our experiments.

The MMS dataset was divided into two parts. One part contains 80% for learning and the other part contains the remaining 20% for the test. According to the confusion matrix of the Text-Image Fusion Model described in Table 2, the high correct classification is when spam categories are predicted as spam classes. Also, the model has identified spam MMS with low false-negative and false-positive rates.

To verify the performance of MMS-FS, our analysis uses different datasets to compare it with well-performing models. For the image-based dataset, [6] (C), [8] (D), [2] (E), and MMS-FS were used for comparison. For the text-based dataset, [5] (A), Smishing Detector [3] (B), (K) [4], and the MMS-FS model were used for comparisons. For the MMS dataset, the result of MMS-FS was displayed to verify its performance.

As shown in Table 3, in unimodal experiments text-based, the results exposed that our model (MMS-FS) gives the best results compared with (B) and (K), and outperforms (A) in terms of f1-score, recall, and precision. For unimodal experiments image-based, the accuracy of MMS-FS is 97.98%. It gives the best accuracy, recall, f1-score, and precision compared with (C) and (E). For the multimodal experiments, the overall result shows that our filtering system is efficient with an accuracy of 98.56%. We conclude that MMS-FS implements the MMS spam filtering function well, whether the spam content is hidden in the text, in the image, or hidden in the text and image.

A ROC (Receiver Operating Characteristic) chart is the test's sensitivity averaged over its specificity or vice versa. So, we utilize a ROC chart to better clarify the effectiveness of the MMS-FS. In Fig. 7a–c illustrates the ROC charts for the text-based, image-based, and MMS datasets. We can note that the AUC indicators of the proposed system are higher than 0.97. It appears from these results that the accuracy of MMS-FS on the three datasets is very good for spam filtering.

Table 3 Experimental comparisons between different datasets

Dataset	Model	Accuracy (%)	Precision (%)	Recall (%)	F1-score (%)
Text-based dataset	(A)	98	95	93	93.96
	(B)	96.29	96	95.5	95
	(K)	96.40	96	96.25	94.47
	MMS-FS	96.79	**96.79**	**96.79**	**97.23**
Image-based dataset	(C)	97	–	–	96.2
	(D)	98	–	–	–
	(E)	92.1	92	92	92
	MMS-FS	97.98	97.98	97.98	97.94
MMS dataset	MMS-FS	**98.56**	**98.56**	**98.56**	**98.56**

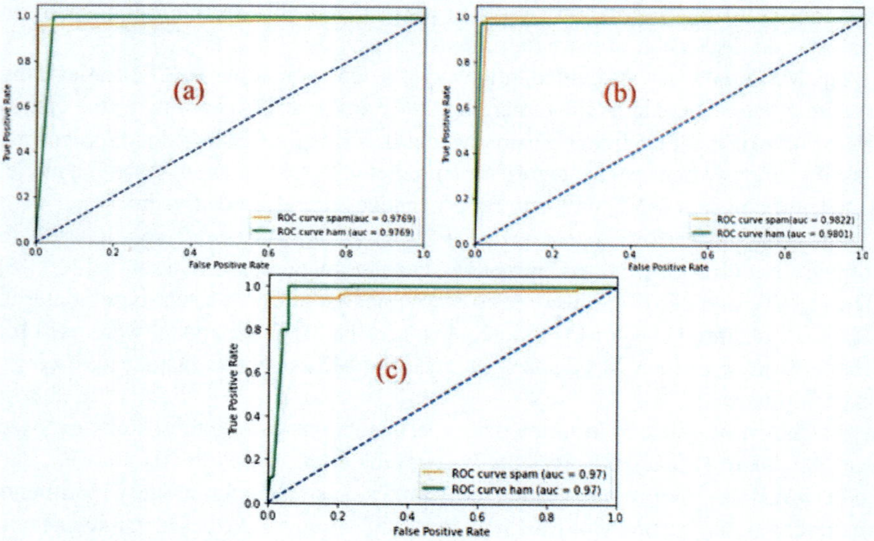

Fig. 7 ROC curve analysis results

5 Conclusion

Our study focused on an important issue of messaging spam detection. We proposed MMS-FS, a novel Multi-Modal system based on LSTM and CNN models. The performance evaluation of this system shows a rate of 0.9798 for the image-based dataset, 0.9679 for the text-based dataset, and 0.9856 for the MMS dataset. Also, by comparison with the most similar and relevant research, we notice an improvement in recall, f1-score, accuracy, and precision of smashing detection between 1 to 5%. Therefore, we can conclude from the experimental evaluation that our system can effectively detect spam content in MMS, whether it is hidden in the text, in the image, or hidden in the text and image. Other social media data can be processed using the proposed method. For better performance results, it may be necessary to increase the collection of social media text samples during the training process, and to measure the maximum performance of our architecture, we can use other CNN architectures or Deep Learning methods. In the future, we plan to incorporate more techniques into MMS-FS to prevent more intelligent attacks. Also, using transformer models, we can improve the detection performance of multimodal architectures by enriching both image and text representation.

References

1. MMS Marketing—What Marketers Need to Know. Tatango—SMS Marketing Software, Oct. 06, 2020. https://www.tatango.com/blog/mms-marketing-what-marketers-need-to-know/. Accessed 30 May 30
2. H. Yang, Q. Liu, S. Zhou, Y. Luo, A spam filtering method based on multi-modal fusion. Appl. Sci. **9**(6), 1152 (2019). https://doi.org/10.3390/app9061152
3. S. Mishra, D. Soni, Smishing Detector: a security model to detect smishing through SMS content analysis and URL behavior analysis. Future Gener. Comput. Syst. **108**, 803–815 (2020). https://doi.org/10.1016/j.future.2020.03.021
4. G. Sonowal, K.S. Kuppusamy, SmiDCA: an anti-smishing model with machine learning approach. Comput. J. **61**(8), 1143–1157 (2018). https://doi.org/10.1093/comjnl/bxy039
5. A. Annadatha, M. Stamp, Image spam analysis and detection. J. Comput. Virol. Hacking Tech. **14**(1), 39–52 (2018). https://doi.org/10.1007/s11416-016-0287-x
6. Y. Gao et al., Image spam hunter, in *2008 IEEE International Conference on Acoustics, Speech and Signal Processing*, Las Vegas, NV, USA (2008), pp. 1765–1768. https://doi.org/10.1109/ICASSP.2008.4517972
7. S. Srinivasan et al., Deep convolutional neural network based image spam classification, in *2020 6th Conference on Data Science and Machine Learning Applications (CDMA)*, Riyadh, Saudi Arabia (2020), pp. 112–117. https://doi.org/10.1109/CDMA47397.2020.00025
8. J. Bouvrie, *Notes on Convolutional Neural Networks*, p. 8 (2006)
9. G. Jain, M. Sharma, B. Agarwal, Optimizing semantic LSTM for spam detection. Int. J. Inf. Technol. **11**(2), 239–250 (2019). https://doi.org/10.1007/s41870-018-0157-5
10. T.A. Almeida, J.M. Gómez, A. Yamakami, Contributions to the study of SMS spam filtering: new collection and results, in *Proceedings of the 11th ACM Symposium on Document Engineering* (2011), pp. 259–262

Optimized Neural Network for Evaluation Cisplatin Role in Neoplastic Treatment

Ahmed T. Sahlol, Ahmed A. Ewees, and Yasmine S. Moemen

Abstract Cisplatin causes genetic damage to the DNA; however, it is usually used against diverse cancers. The gene expression of the treated cells should be evaluated to measure such damage. This paper proposes an optimized neural network for evaluating Cisplatin efficiency as a chemotherapeutic drug optimized by the Multi-Verse Optimizer (MVO). The proposed approach starts with updating the weights and biases of the network until reaching the optimal value, which improves prediction accuracy. Publicly available data from the Health and Environmental Sciences Institute (HESI) was used to validate Cisplatin's impact on gene expression levels in cancer cells. The proposed optimization approaches have achieved the lowest MSE in all experiments, which means more efficiency than the other classical models. Even when comparing our optimization models with other models, it achieved the highest performance among them.

Keywords Neural Network (NN) · Multi-verse Optimizer (MVO) · Cisplatin efficiency · Chemotherapeutic drug

1 Introduction

Cisplatin is an alkylating drug; it acts as a metal salt. Cisplatin was applied for various types of cancer [1] like prostate, testicular, breast bladder, ovarian, head, neck, lung, esophageal, cervical, breast, stomach, and prostate cancers, besides the treatment of Hodgkin's, non-Hodgkin's lymphomas, Sarcomas, Neuroblastoma, Melanoma, Multiple Myeloma, and Mesothelioma.

A. T. Sahlol · A. A. Ewees (✉)
Computer department, Damietta University, Damietta, Egypt
e-mail: a.ewees@hotmail.com

Y. S. Moemen
Clinical Pathology Department, National Liver Institute, Menoufia University, Shebin El-Kom 32511, Egypt

M. Abd Elaziz et al. (eds.), *International Conference on Artificial Intelligence Science and Applications (CAISA)*, Advances in Intelligent Systems and Computing 1441,
https://doi.org/10.1007/978-3-031-28106-8_10

Certain precautions must be applied when Cisplatin is administered through vein [2]. Unfortunately, the current drug for cancer cannot recognize the cancerous cell from the normal cell. This drug harms the normal cell in different areas of the body as blood cells, mouth, stomach, and bowel; such drugs cause hair fall, blood counts decrease, mouth inflammations, vomiting, and diarrhea [2]. The genetic damage of Cisplatin was mentioned before by measuring genes that identify genotoxic compounds [3, 4].

Cisplatin is known as cisplatinum and Cisdiamminedichloroplatinum (II); it was approved in 1978, and its biological activity was mentioned after 125 years of the first synthesis in 1845 [5]. It is genotoxicity because it interacts with human DNA [6, 7], which inhibits DNA repair mechanisms, causes DNA damage, and leads to apoptosis in cancer cells [8]. Also, other actions happen, such as drug resistance, the adverse effect on kidneys, vulnerability to infections, gastritis, bleeding, and hearing loss, especially in younger patients [1]. To treat such issues, the gene expression technique was used to identify the drug response. Several studies used TK6 cells as a genetic marker to define compounds as genotoxic or nongenotoxic [9, 10]. TK6 cells were treated with different measures of cisplatin concentrations for 4h and collected 0h, 4h, 20h, or increased post-treatment period for each study procedure. Gene expression output data were inspected and estimated for each study condition.

The main target of the microarray investigation technique is to personalize the chemotherapy, expect drug productivity, and procedure [11]. As the microarray technique vulnerability to gene expression is not accurate enough [4, 12].

This study aims to detect better gene levels involved in the chemosensitivity of Cisplatin by predicting the over-expression of specific genes, which is considered a genetic marker of adverse effects. In this work, we aim to predict the gene expressions (about 40 Genes) based on the study period and the active Cisplatin concentrations using a neural network optimized by the multi-verse optimizer (MVO) method. The MVO is an optimization algorithm that showed promising results in several previous studies in solving many optimization problems such as [13–17]. In addition, several optimization algorithms were applied successfully to solve different problems, such as image segmentation [18], classification of white blood cell leukemia [19], mortality incidence in Tilapia fish [20].

The rest of the paper is arranged as follows: Sect. 2 shows the related works. Section 3 presents data characteristics. Section 4 discusses the MVO algorithm. Section 5 presents the proposed cisplatin prediction approach. Results with discussions are described in Sect. 6. Finally, conclusions and future work are provided in Sect. 7.

2 Related Works

The main purpose of this paper is to present a new approach for predicting gene expressions as a result of cisplatin doses. The proposed approach is based on NN optimized by Multi-Verse Optimizer (MVO). In this section, we outline some state of the arts that deployed NN in medical systems. Some of them used it as a classifier, others used it as predictors.

In Pasomsub et al. [21], a phenotypic drug resistance prediction model was built by a neural network. The model was applied to forecasting the HIV-1 resistance phenotype from the genotype. Drug resistance was accurately predicted by the proposed NN model and generalized for individual HIV-1 subtypes. While in [22], it was applied to the approximation of genetic function and neural network to model the activities of a series of HIV reverse transcriptase inhibitor TIBO derivatives. Multivariate image analysis was applied to the quantitative structure-activity relationship. Analysis-adaptive Neuro-Fuzzy Inference Systems (ANFIS) adapted by PCA were applied to the same set of compounds. The proposed ANFIS methodology successfully handled Quantitative Structure-Activity Relationship (QSAR) problems.

In addition, Dechao Wang et al. [23] used a neural network to predict virological response combined with HIV therapy. The combination of both; HIV genotype therapy and other clinical information were deployed to predict virological using neural networks. The neural network showed higher performance than tree-based algorithms and SVM.

Nelwamondo et al. [24] studied missing data using an industrial power plant and the industrial winding process by developing neural networks and expectation-maximization techniques. Two estimations of missing data approaches were followed based on maximum likelihood (ML) and expectation maximization (EM). A hybrid model was built based on a genetic algorithm and neural network (NN-GA). They recommended the hybrid (NN-GA) because of its inherent nonlinear relationships between the trial variables. On the contrary, the (NN-EM) model can be recommended and performed better if there is little interdependent between the input variables.

Another application by Betechuoh et al. [25] was proposed to achieve adaptive control of HIV status by a neural network. The genetic algorithm was used to search the space to find the optimal number of hidden units of feed-forward and inverse neural networks. The performance of the learning level would be high if only the HIV status were known. The authors of [26] improved the neural network using the sine-cosine algorithm to improve the liver enzymes prediction on fish farmed on nano-selenite. The improved version of NN achieved better results than the original NN.

In this context, in [27], a hybrid traditional neural network and genetic algorithm for Indian diabetes classification were presented. Two methods were used to initialize the Back Propagation neural network weights: A decision tree and a genetic algorithm. The optimal weights were achieved by using a hybrid Genetic algorithm. It was capable of effectively exploring large search spaces. It also showed a substantial improvement in classification accuracy. It also showed better results when used as a feature selector to improve the classification accuracy of the neural networks.

Finally, Dheeba et al. [28] proposed a hybrid classification method for detecting breast abnormalities in mammograms by applying a particle swarm with a wavelet neural network. They extracted texture energy measures from the mammograms of a clinical database of 216 mammograms, then classified the abnormal regions by applying machine learning algorithms. The model's performance was validated using a ROC curve, which calculated the trade-offs between the sensitivity and specificity of the proposed system. The results showed that the area under the ROC curve of the model reached 0.96853 with a sensitivity and specificity of 94.167 and 92.105%, respectively.

3 Data Characteristics

Genotoxicity of various compounds such as Cisplatin, sodium chloride, and Taxol was proven in various research as markers of DNA damage. Data was collected from the Health and Environmental Sciences Institute (HESI) [5]. In this study, Cisplatin was only used as a marker of DNA damage. In the previous studies, two genotoxicity testing [L5178Y mouse lymphoma and human thymine kinase (TK) 6 cells] were performed. Those tests showed low levels of p53 and high levels of the human TK6 cells in mouse lymphoma cells. In all studies, the gene expression profiles of TK6 cells were also used for about 4 and 24 h. This was performed to standardize measurement, handling at least 31 genes expressed over all the studies out of the list of 44 genes. To measure the level of gene expression of a range between (42-49) genes, a comparison was made between various concentrations of anti-neoplastic Cisplatin and different study duration. The study duration was four hours, twenty-six hours, twenty-eight hours, thirty-one hours, and forty-eight hours as an upper limit. The dosage of Cisplatin was categorized as high, medium, and low, as is shown in Table 1.

Table 1 Experiment settings

Study No.	Time of exposure (hours)	Concentrations (μ g)	Expressed gene
1	28,31	Low =1	37
	28,31,48	Medium = 10	
	28,31,48	High = 30	
2	28,31,48	Low = 1	42
	28,31,48	Medium = 10	
	28,31,48	High = 30	
3	28,48	Low = 0.1	44
	28,48	High = 100	
4	28,31,48	Low = 0.1	41
	28,31,48	Medium = 1	
	28,31,48	High = 10	
5	26,28,31,48	Low = 1	31
	26,28,31,48	Medium = 10	
	28,31,48	High = 30	

In previous works, study [29] 008-00004-0006-000-8, Study [30] 008-00004-0004-000-6, Study [31] 008-00004-0002-000-4, Study [32] 008-00004-0007-000-9 and Study [33], different concentrations of Cisplatin with multiple concentrations were dissolved in Dimethylsulfoxide (DMSO). The concentration of DMSO in the cell culture medium did not exceed 1%. Test solutions were prepared immediately before the usage, and each study's treatment period was quite similar.

GADD45 family exhibited strong relations with the amount of DNA-platinum adducts, its up-regulation reached 14 folds [11], this also accompanied by increased mRNA levels of the p53 target genes MDM2, TP53I3, and PPM1D [3]. Datasets 4 and 5 have bad predictions when compared to datasets 1, 2 and 3. The current study target such work, to give more integrity and improve interpretations.

4 Multi-verse Optimizer

MVO is a type of meta-heuristic method. It simulates the multi-verse theory in physics [34]. This theory contains three types of holes (white–black–worm), and the MVO algorithm simulates the transferring objects and interaction between these holes.

4.1 Inspiration

Mirjalili et al. [34] commented that the multi-verse concept means there is more than one universe in addition to our universe. These universes, based on multi-verse theory, can communicate with each other. The multi-verse theory assumes that a white hole is formed after the parallel universes' collision, whereas a black hole attracts anything by its extremely high gravitational force. Wormholes connect different parts of the universe. The objects are passed between various universes by white/black hole tubes. When a white/black tube is settled among two universes, the universe with a great inflation rate is recognized to have a white hole; otherwise, a black hole can appear. After that, objects are allowed to transfer from the white to the black holes.

Mechanism

MVO algorithm explores search spaces based on the white/black holes' concept to improve the ability to exploit the search spaces. It begins by creating random numbers of universes. In each loop, objects are transferred between white and black holes based on the value of the objective function till meeting the end criteria. MVO represents the solution by a universe, whereas the object is the solution's variable.

In MVO [34], a solution is represented by a universe, and the solution's variable is formed as an object in the universe. The value of the fitness function is represented as an inflation rate.

The following equations represent MVO mathematically.

$$U = \begin{bmatrix} x_1^1 \; x_1^2 \; \ldots \; x_1^d \\ x_2 \; x_2^2 \; \ldots \; x_2^d \\ \vdots \;\; \vdots \;\; \vdots \;\; \vdots \\ x_n^1 \; x_n^2 \; \ldots \; x_n^d \end{bmatrix} \tag{1}$$

$$x_i^j = \begin{cases} x_k^j & if r_1 < NI(U_i) \\ x_i^j & otherwise \end{cases} \tag{2}$$

where U is a universe. d defines the problem's dimension. N is the universes' number. r_1 in range [0,1], U_i defines universe ith. $NI(U_i)$ denotes the normalized objective value of i.

The wormholes work to improve the value of the objective function by updating the best universes' objects as follows:

$$x_i^j = \begin{cases} \begin{cases} X_j^b + TDR \times ((ub_j - lb_j) \times r_4 + lb_j) & r_3 < 0.5 \\ X_j^b - TDR \times ((ub_j - lb_j) \times r_4 + lb_j) & r_3 \geq 0.5 \end{cases} & r_2 < WEP \\ x_i^j & r_2 \geq WEP \end{cases} \tag{3}$$

Where X_j^b defines the best solution. The lower and upper limit of jth variable are defined by lb_j and ub_j. All versions of r are random numbers in the range [0, 1]. TDR is called the traveling distance rate; it is a coefficient used to locate the best universe to allow a wormhole to move to it. TDR is calculated by (4). WEP is the wormhole existence probability. It is applied to increase the effectiveness of the exploitation phase. WEP is calculated by (5).

$$TDR = 1 - \frac{i^{1/p}}{I^{1/p}} \tag{4}$$

$$WEP = WEP_{min} + t \times \left(\frac{WEP_{\max} - WEP_{\min}}{T} \right) \tag{5}$$

where i is the current iteration of I iterations. p is a constant equal 6. $WEP_{\min}$ and $WEP_{\max}$ equal 0.2 and 1, respectively.

5 Proposed Neural Network Optimized by MVO

Neural Network (NN) is mainly used for pattern classification; however, in other works, it is used for regression (prediction). Prediction is concerned with modeling the relationship between the current and previous data, then generating and forecasting future trends. While classification is concerned with dividing data into categories.

In practice, running too many training iterations is required until the network error reaches a satisfactory level (low prediction error). Training has to be terminated once the performance of the NN becomes acceptable. In this paper, we apply MVO for optimizing neural network training by updating their weights and biases. So, instead of performing hundreds of iterations manually until reaching the optimum values of neural weights and biases, MVO searches the space effectively to find their optimum values.

MVO consists of a set of particles, which represents a solution. Each particle changes its position according to its neighbors and its own experience. It is widely common that using optimization algorithms makes the training phase very slow [35] and may not converge to a single global minimum because there exist many local minima. However, randomization-based methods fix this problem by randomly adjusting network parameters. MVO starts by creating a random population of candidate solutions for the optimization problem. Each solution is a vector of length N (where N is the number of all potential solutions to the problem). Figure 1 represents the NN optimization mechanism.

Therefore, through iterations, MVO parameters work on approximating probable position of its target. We believe that neural network is sufficient in dealing with such a wide range of data especially in biological fields based on our previous work [26]. This work applied classification accuracy to evaluate all solutions (select the best weight and bias value) to achieve fitness function. Figure 2 illustrates the overall optimized-NN approach.

Objective function

The proposed approach adopted Mean Square Error (MSE) as a cost function of MVO to evaluate the solutions. The evaluation is performed by representing the existing solution as weights and biases, then passing it to the NN. Finally, MSE is computed. This technique is repeated until reaching the max number of iterations.

Fig. 1 NN optimization mechanism

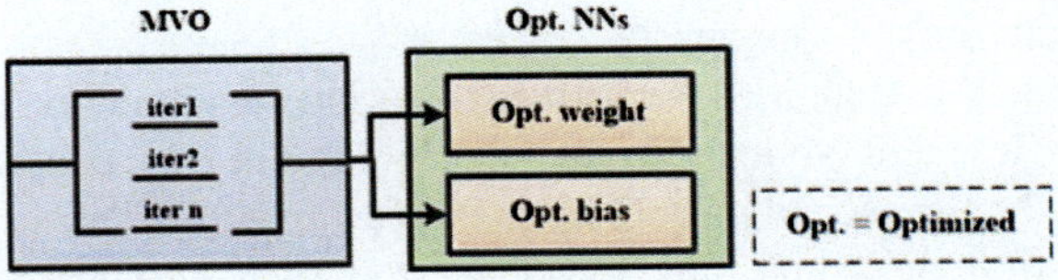

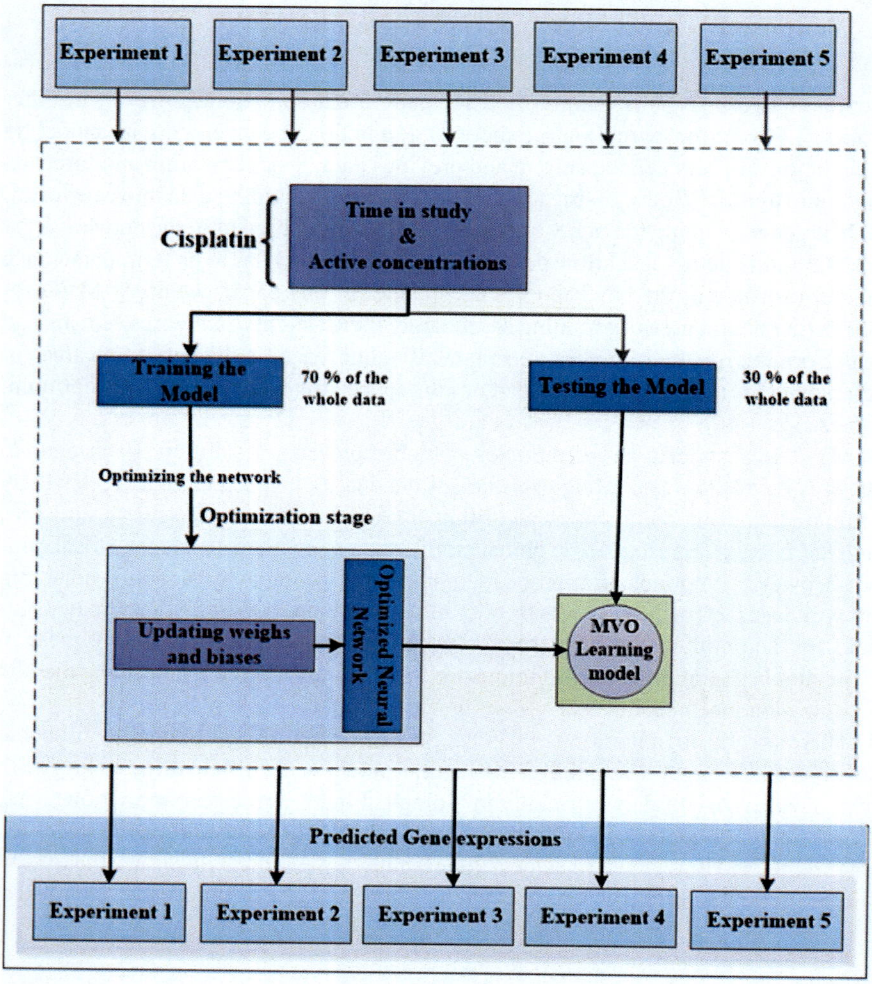

Fig. 2 Proposed NN-MVO prediction approach

Then, the best solution is passed to the NN to be used in the testing phase. MSE is computed as in Eq. 6.

$$MSE = \frac{1}{n} \Sigma_1^n (y_i - x_i)^2) \tag{6}$$

where y_i defines the ith predicted values. x_i defines the actual values. n defines the number of samples.

We adopt MSE again for measuring the prediction performance. It measures how close a fitted line is to each data point. For each one, the distance is calculated from the corresponding value of the fitted curve line. Then, we subtract the predicted value from the actual value to get an error. Then, we square that error and add up all other errors. Finally, we get the mean of all errors. Achieving a smaller MSE value is a key target in such problems, which means that the predicted data is close to the actual data.

As seen in Fig. 2, The inputs of the NN-MVO model include the study duration (26 H, 28 H, 31 H, or 48 H) and active concentrations (0.1 ug/ml, 1 ug/ml, 10 ug/ml, 30 ug/ml or 50 ug/ml). The samples were divided randomly based on 70, 15, and 15% for training, validation, and testing. The testing set was used Only for testing the model's efficiency. There is no overlap between any of the sets.

In this paper, the dataset was challenging because it contains only two inputs and multi-outputs (about 40 outputs).

6 Experiments and Discussions

This section presents the parameter settings, results of experiments, and discussions for the testing samples.

6.1 Parameter Settings

The experiment uses nine hidden layers, 1000 max epochs, and Levenberg-Marquardt as a training function. These parameters were chosen after many experimental tries, and they achieved the best performance.

The rest of the parameters for NN and MVO are outlined in Table 2.

All experiments were built in "MATLAB 2014b" on "Windows 10" with "4 GHz" of memory. Data preprocessing and normalization are important steps in data analysis. So, we scale the data to be in the interval of [0, 1] as that might better influence a prediction performance based on [36], who used the same technique.

To investigate the relationship between the number of samples and the efficiency of the prediction approach for each of the five datasets, MSE was measured for each of them, Fig. 3.

Table 2 Parameter settings

Parameter	Value
Domain	[0, 1]
N	30
Max iteration of	200
Number of iterations (runs)	20
MVO	$WEP_{\min} = 0.2$, $WEP_{\max} = 1$

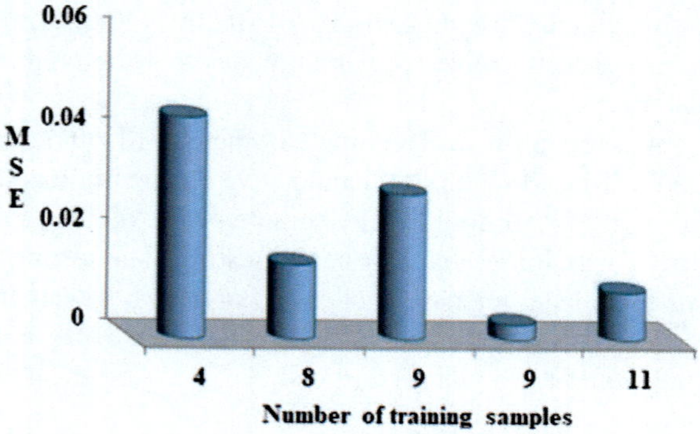

Fig. 3 Relationship between number of training samples and MSE

Defining DNA reactive from non-DNA reactive genotoxic pathways might need fewer genes [37]. A comparative study performed an analysis of gene expression evaluations on about 40 genes of different studies, where Cisplatin is used as a test compound for Cytotoxicity [3] because Cisplatin forms directly across a link with the DNA of the cell.

Several experiments on each dataset were performed to test the efficiency of the proposed approach. Each MVO iteration is called a round. MVO performed five rounds on each dataset to predict the expression of about 40 genes.

6.2 *Comparative Analysis*

Optimization approaches are iterative-based as they try to solve problems by sampling a probability distribution over the search space. Then, this probability distribution is modified using better ones among constructed solutions. So, the main purpose is to sample the problem solution for the areas of research space with a high potential for solutions. For the first three datasets, it is noticed that MVO achieved an acceptable prediction rate which proves to have high mobility and degree of exploration. Besides, one of the most interesting things is the use of $g*$ (the current global best solution), which seems to be strongly selective as it supports the convergence by drawing toward the current best $g*$. At the same time, it might lead to uncorrected convergence, which may not be considered the true optimal solution.

The last two datasets showed lower efficiency (compared to the first three). We think this is also reasonable, as the data were unbalanced and the gap between gene values was high. In this paper, it was challenging that the studied problem (gene expression) has only two inputs, while at the same time, the proposed approach has to predict 40 gene expression values. By taking advantage of swarm intelligence, the problem space (gene expressions) was explored efficiently to find the optimal solution (the best values of the weight and the bias of the neural network). So accordingly, we were able to achieve better convergence, which reflected in achieving a high-quality gene expression prediction.

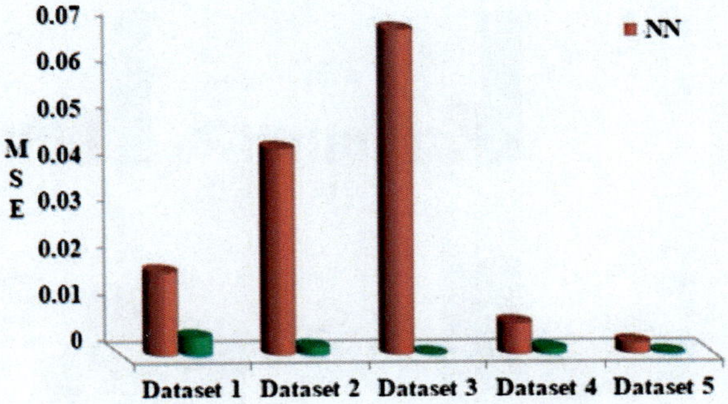

Fig. 4 Comparison between the NN and NN-MVO

Table 3 Results of the proposed method and the compared methods

Dataset	NN	NN-PSO	NN-GA	NN-GWO	NN-MVO
1	0.0239	0.0086	0.0044	0.0079	0.0040
2	0.0421	0.0021	0.0021	0.0018	0.0017
3	0.0015	0.0009	0.0729	0.0091	0.0002
4	0.0016	0.0017	0.0023	0.0026	0.0012
5	0.0014	0.0027	0.0008	0.0024	0.0002

To validate the used optimization approach, MSE for five MVO rounds in each dataset were calculated and compared to those achieved by un-optimized NN (Neural Network with Back Propagation "NN"). Figure 4 shows a comparison between the NN and NN-MVO.

In NN, the gradient descent function searches for the minimum error in the weight space. Since this method requires the computation of the gradient of the error function at each iteration. The activation function used in this work is the most popular; the Sigmoid function.

It is obvious that NN-MVO achieved better results than non-optimized NN (NN-BP) in most datasets.

The proposed prediction approach was also compared to other recent optimization algorithms (that optimized neural network weight and biases) to investigate the ability of the proposed approach among other optimization algorithms. These algorithms are Particle Swarm Optimization (PSO), Grey Wolf Optimization (GWO), and genetic algorithm (GA). All experiments used the default parameter setting for the optimizers. Experimental results are shown in Table 3 and Fig. 5.

The proposed approach (NN-MVO) proved to achieve better accuracy in terms of minimizing the prediction errors between the produced and the actual gene expressions in most experiments (datasets). It achieved low MSE values when applied to the five datasets, which means that it is reliable and can work efficiently with multi-output prediction problems.

Defining genotoxic mechanisms with currently known assays is done using many experimental studies with unpredictable results, which is considered challenging

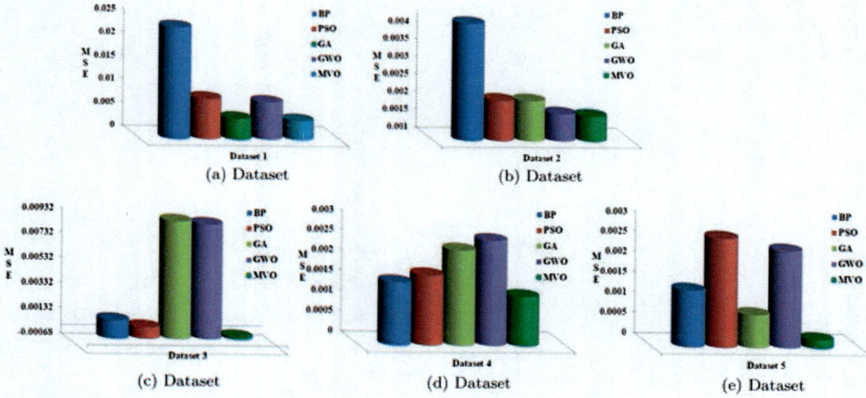

(a) Dataset (b) Dataset (c) Dataset (d) Dataset (e) Dataset

Fig. 5 Comparison with other NN based optimization algorithms

for novel medicines' approval. So, another mechanism-based testing technique is urgently needed [3]. While microarray is considered an important tool for evaluating gene expression changes, it is still not sensitive enough to signify genotoxicity [11]. The current paper aimed to improve results extracted from previous studies, which might be, consequently, lower cost compared to other lab techniques.

7 Conclusion

This paper aims to define potential differences in the gene expression pattern of Cisplatin activity in the case of lymphomas through a comparison of the level of gene expression involved in Cisplatin activity as described by different concentrations and different studies. The proposed prediction approach is based on Neural Network Optimized by the multi-Verse Optimizer (NN-MVO). The proposed approach aims at improving gene expression prediction accuracy by iteratively updating NN weights and biases. Achieving the best ones enables NN to achieve higher prediction accuracy efficiently. Five datasets were used in this work; each has only two inputs and about 40 outputs (representing genes). Several rounds of MVO were achieved to validate the performance of the proposed approach. NN-MVO shows acceptable prediction accuracy, while it also shows differential results. The proposed approach achieved high prediction performance as it could search the problem space efficiently until reaching the optimum weight and bias values. When comparing the results with those performed by the backpropagation neural network, it was clear that the proposed approach achieved better prediction performance. Moreover, compared to other recent optimization algorithms, NN-MVO outcomes them in most experiments.

References

1. S. Dasari, P.B. Tchounwou, Cisplatin in cancer therapy: molecular mechanisms of action. Eur. J. Pharmacol. **740**, 364–378 (2014)
2. R. Oun, Y.E. Moussa, N.J. Wheate, The side effects of platinum-based chemotherapy drugs: a review for chemists. Dalton Trans. **47**(19), 6645–6653 (2018)
3. H. Ellinger-Ziegelbauer, J.M. Fostel, C. Aruga, D. Bauer, E. Boitier, S. Deng, D. Dickinson, A.-C. Le Fevre, A.J. Fornace Jr, O. Grenet et al., Characterization and interlaboratory comparison of a gene expression signature for differentiating genotoxic mechanisms. Toxicol. Sci. **110**(2), 341–352 (2009)
4. R.K. Newton, M. Aardema, J. Aubrecht, The utility of DNA microarrays for characterizing genotoxicity. Environ. Health Perspect. **112**(4), 420–422 (2004)
5. S.M. Cohen, S.J. Lippard, Cisplatin from DNA damage to cancer chemotherapy. Progress Nucleic Acid Res. Mol. Biol. **67**, 93–130 (2001)
6. J. Reedijk, Why does cisplatin reach guanine-n7 with competing s-donor ligands available in the cell? Chem. Rev. **99**(9), 2499–2510 (1999)
7. G.L. Cohen, J.A. Ledner, W.R. Bauer, H. Michael Ushay, C. Caravana, S.J. Lippard, Sequence dependent binding of cis-dichlorodiammineplatinum (ii) to DNA. J. Am. Chem. Soc. **102**(7), 2487–2488 (1980)
8. L. Kelland, The resurgence of platinum-based cancer chemotherapy. Nat. Rev. Cancer **7**(8), 573 (2007)
9. J.K. Buick, I. Moffat, A. Williams, C.D. Swartz, L. Recio, D.R. Hyduke, H.-H. Li, A.J. Fornace Jr, J. Aubrecht, C.L. Yauk, Integration of metabolic activation with a predictive toxicogenomics signature to classify genotoxic versus nongenotoxic chemicals in human tk 6 cells. Environ. Mol. Mutagen. **56**(6), 520–534 (2015)
10. A. Williams, J.K. Buick, I. Moffat, C.D. Swartz, L. Recio, D.R. Hyduke, H.-H. Li, A.J. Fornace, J. Aubrecht, C.L. Yauk, A predictive toxicogenomics signature to classify genotoxic versus non-genotoxic chemicals in human tk6 cells. Data Brief **5**, 77–83 (2015)
11. M. Sakamoto, A. Kondo, K. Kawasaki, T. Goto, H. Sakamoto, K. Miyake, Y. Koyamatsu, T. Akiya, H. Iwabuchi, T. Muroya et al., Analysis of gene expression profiles associated with cisplatin resistance in human ovarian cancer cell lines and tissues using cdna microarray. Hum. Cell **14**(4), 305–315 (2001)
12. H.-H. Li, R. Chen, D.R. Hyduke, A. Williams, R. Frötschl, H. Ellinger-Ziegelbauer, R. O'Lone, C.L. Yauk, J. Aubrecht, A.J. Fornace (2017) Development and validation of a high-throughput transcriptomic biomarker to address 21st century genetic toxicology needs, in *Proceedings of the National Academy of Sciences* (2017), pp. 201714109
13. H. Faris, M.A. Hassonah, A.-Z. Ala'M, S. Mirjalili, I. Aljarah, A multi-verse optimizer approach for feature selection and optimizing SVM parameters based on a robust system architecture. Neural Comput Appl **30**(8), 2355–2369 (2018)
14. A.A. Ewees, M.A. El Aziz, A.E. Hassanien, Chaotic multi-verse optimizer-based feature selection. Neural Comput. Appl. 1–16 (2017)
15. H. Faris, I. Aljarah, S. Mirjalili, Training feedforward neural networks using multi-verse optimizer for binary classification problems. Appl. Intell. **45**(2), 322–332 (2016)
16. A.A. Ewees, M.A. Elaziz, Performance analysis of chaotic multi-verse Harris hawks optimization: a case study on solving engineering problems. Eng. Appl. Artif. Intelli. **88**, 103370 (2020)
17. M.A. Elaziz, D. Oliva, A.A. Ewees, S. Xiong, Multi-level thresholding-based grey scale image segmentation using multi-objective multi-verse optimizer. Expert Syst. Appl. **125**, 112–129 (2019)
18. A.A. Ewees, L. Abualigah, D. Yousri, A.T. Sahlol, M.A. Al-qaness, S. Alshathri, M.A. Elaziz, Modified artificial ecosystem-based optimization for multilevel thresholding image segmentation. Mathematics **9**(19), 2363 (2021)
19. A.T. Sahlol, P. Kollmannsberger, A.A. Ewees, Efficient classification of white blood cell leukemia with improved swarm optimization of deep features. Sci. Rep. **10**(1), 1–11 (2020)

20. A.A. Ewees, A.A. Hemedan, A.E. Hassanien, A.T. Sahlol, Optimized support vector machines for unveiling mortality incidence in tilapia fish. Ain Shams Eng. J. **12**(3), 3081–3090 (2021)
21. E. Pasomsub, C. Sukasem, S. Sungkanuparph, B. Kijsirikul, W. Chantratita et al., The application of artificial neural networks for phenotypic drug resistance prediction: evaluation and comparison with other interpretation systems. Jpn. J. Infect. Dis. **63**(2), 87–94 (2010)
22. M. Goodarzi, M.P. Freitas, Mia–GSAR coupled to principal component analysis-adaptive neuro-fuzzy inference systems (PCA–ANFIS) for the modeling of the anti-HIV reverse transcriptase activities of TIBO derivatives. Eur. J. Med. Chem. **45**(4), 1352–1358 (2010)
23. D. Wang, B. Larder, A. Revell, J. Montaner, R. Harrigan, F. De Wolf, J. Lange, S. Wegner, L. Ruiz, M.J. Pérez-Elías et al., A comparison of three computational modelling methods for the prediction of virological response to combination HIV therapy. Artif. Intell. Med. **47**(1), 63–74 (2009)
24. F.V. Nelwamondo, S. Mohamed, T. Marwala, Missing data: a comparison of neural network and expectation maximization techniques. Curr. Sci. 1514–1521 (2007)
25. B.L. Betechuoh, T. Marwala, T. Tettey, Using inverse neural networks for HIV adaptive control. Int. J. Comput. Intell. Res. **3**(1), 11–15 (2007)
26. A.T. Sahlol, A.A. Ewees, A.M. Hemdan, A.E. Hassanien, Training feedforward neural networks using sine-cosine algorithm to improve the prediction of liver enzymes on fish farmed on nano-selenite, in *2016 12th International Computer Engineering Conference (ICENCO)* (IEEE, 2016), pp 35–40
27. A.G. Karegowda, A.S. Manjunath, M.A. Jayaram, Application of genetic algorithm optimized neural network connection weights for medical diagnosis of pima Indians diabetes. Int. J. Soft Comput. **2**(2):15–23 (2011)
28. J. Dheeba, N. Albert Singh, S. Tamil Selvi, Computer-aided detection of breast cancer on mammograms: a swarm intelligence optimized wavelet neural network approach. J. Biomed. Inform. **49**, 45–52 (2014)
29. J. Aubrecht, 4_tk6_cisplatin-etoposide-nacl-taxol, health and environmental sciences institute, genetic toxicology sapiens human. http://tools.niehs.nih.gov/cebs3/ui/?study=008-00004-0009-000-1 (2004). Accessed 1 May 2018
30. J. Aubrecht, 4_tk6_cisplatin-etoposide-nacl-taxol, health and environmental sciences institute, genetic toxicology sapiens human. http://tools.niehs.nih.gov/cebs3/ui/?study=008-00004-0002-000-4 (2004). Accessed 1 May 2018
31. J. Aubrecht, 4_tk6_cisplatin-etoposide-nacl-taxol, health and environmental sciences institute, genetic toxicology sapiens human. http://tools.niehs.nih.gov/cebs3/ui/?study=008-00004-0004-000-6 (2004). Accessed 1 May 2018
32. J. Aubrecht, 4_tk6_cisplatin-etoposide-nacl-taxol, health and environmental sciences institute, genetic toxicology sapiens human. http://tools.niehs.nih.gov/cebs3/ui/?study=008-00004-0006-000-8 (2004). Accessed 1 May 2018
33. J. Aubrecht, 4_tk6_cisplatin-etoposide-nacl-taxol, health and environmental sciences institute, genetic toxicology sapiens human. http://tools.niehs.nih.gov/cebs3/ui/?study=008-00004-0007-000-9 (2004). Accessed 1 May 2018
34. S. Mirjalili, S.M. Mirjalili, A. Hatamlou, Multi-verse optimizer: a nature-inspired algorithm for global optimization. Neural Comput. Appl. **27**(2), 495–513 (2016)
35. R.A. Jacobs, Increased rates of convergence through learning rate adaptation. Neural Netw. **1**(4), 295–307 (1988)
36. A.M. Abdeldaim, A.T. Sahlol, M. Elhoseny, A.E. Hassanien, *Computer-Aided Acute Lymphoblastic Leukemia Diagnosis System Based on Image Analysis* (Springer International Publishing, Cham, 2018), pp. 131–147
37. A.-C. Le Fevre, E. Boitier, J.-P. Marchandeau, A. Sarasin, V. Thybaud, Characterization of DNA reactive and non-DNA reactive anticancer drugs by gene expression profiling. Mutat. Res./Fundam. Mol. Mech. Mutagen. **619**(1), 16–29 (2007)